I0063617

MEGAPROJECTS

RESOURCE BOOK FOR PROJECT PROFESSIONALS

LOUISE HART

Copyright © Louise Hart 2025

All rights reserved. No part of this book may be reprinted or reproduced or utilised in any form or by any electronic, mechanical, or other means, now known or hereafter invented, including photocopying and recording, or in any information or retrieval system, without permission in writing from the publisher.

No AI Training: Without in any way limiting the author's exclusive rights under copyright, any use of this publication to "train" generative artificial intelligence (AI) technologies is expressly prohibited. The author reserves all rights to license uses of this work for generative AI training and development of machine learning language models.

Trademark notice: Product or corporate names may be trademarks or registered trademarks. They are used only for identification and explanation with no intention to infringe.

Cover by Seymour Design

Back cover photo by Hero Shot Photography

ISBN: 978-1-7640338-0-0 (pbk)

978-1-7640338-1-7 (ebk)

For Anthony

CONTENTS

INTRODUCTION

This is a book for megaproject professionals who don't have time to read, which is probably most of you. As a target market for a book, that sounds like a pretty weird choice.

Bear with me.

Some years ago I wrote a book called *Procuring Successful Mega-Projects: How to Establish Major Government Contracts Without Ending up in Court*. It was well-received in the megaproject world (some professionals call it their "bible") and enough people bought it that the publisher brought out a paperback edition in 2024.

Sadly, very few of the people who most needed it, even the ones who bought it, had time to read it. Which has been a bit limiting to my objective of raising standards in megaproject procurement. However much I want to give people the benefit of my 30+ years of experience, I've had to recognise that packing it all into a 100,000-word instruction manual was possibly not the most effective way of passing it on.

So in 2024, having survived 60+ years on this planet without ever having had a Facebook account, I thought it was time to have a go at social media.

LinkedIn seemed to be the place for professionals to hang out. The bulk of posts were variants on "yay me" and "yay team", with much of the remainder devoted to the not-so-subtle art of making money on LinkedIn by teaching other people how to make money on LinkedIn. But there were pockets of people discussing matters of professional interest in (usually) a very civilised way.

So I started posting on LinkedIn. No long articles, no detailed analysis. Usually just 200-300 words, aiming to make one specific point about something related to megaproject procurement, using a real-world example.

Bite-sized chunks.

Guess what? People read them. And engaged with them. And asked for more. Yay me.

But the trouble with social media is, the algorithm decides who gets to see your stuff. I was running into people in the real world who would comment favourably on a post they had seen recently – and would then be staggered to learn there were dozens more that had passed them by completely.

This book is my answer. It's a compilation of my best LinkedIn posts of 2024, plus some bonus ones that never made it onto the web. It's a lot shorter than my previous book, there is no attempt to be comprehensive and it isn't intended to be read cover to cover. Project professionals usually don't have time for that.

I've divided it into parts for your benefit, but wholly without reference to the subject matter: a division just provides a visual break in the text and an excuse to put the kettle on.

Dip into it. It's all still in bite-size chunks. No complex discussions. No footnotes. You can put it by your bedside and read one a day. Bring it to the weekly team meeting and use one as a discussion-starter. Keep it on your desk and try a bit of bibliomancy when your latest megaproject bowls you a googly. Whatever.

Your choice. No pressure.

This is a book, so it doesn't come with a space for comments. But you can find me on LinkedIn.

See you there.

Louise Hart, April 2025

P.S. The thing about keeping stuff bite-sized is that in these mini case studies I've had to be a bit ruthless in selecting facts to make a point. So yes, I do know things were more complicated than I've made them appear. My apologies to anyone who feels I've been unfairly critical.

PART 1

Risk allocated to a contractor ceases to exist

That sounds stupid. Possibly because it is. So why do so many clients behave as if it's true?

It is indeed possible to eliminate certain risks, and it can be a very good thing to do. Every risk eliminated is one less to worry about, and there is generally quite enough to worry about as it is.

But risk allocation is just that – allocation. No matter where the risk is allocated, it still exists.

In the case of a risk allocated to the contractor, the client now doesn't manage that risk directly. Instead, it has to be managed indirectly, by managing the contract. That requires a contract with an appropriate contract management regime, a contract management team with the skills and resources to administer it and the political will to enforce it.

Doesn't always happen.

In the news in 2024 was the example of the Scottish government over the CalMac ferries acquisition – six years late and still counting. Risks notionally allocated to the contractor, amazingly, did not cease to exist. The contract management regime was weakened by a politically-motivated decision to waive a requirement for a Builders Refund Guarantee and the contract was not managed effectively. Instead, in 2019 the failing contractor was nationalised, an extreme measure which has not had a noticeably positive impact on time or cost.

Okay, sometimes the risk does stay with the contractor and the client stays whole while the contractor loses money.

But that's not going to happen if the client isn't willing to invest in properly resourcing the development of a good contract and an effective contract management team.

Famous for the wrong reason

Laufenburg, Switzerland, is separated from Laufenburg, Germany, by the Rhine. There is a bridge joining the two cities, which was opened in December 2004. As the two sides of the bridge were built out from the banks it became apparent that they weren't going to meet in the middle. Missed it by 54cm. Oops.

Yeah, embarrassing.

It seems Switzerland calculates sea level by reference to the Mediterranean and Germany by reference to the North Sea. The levels differ by 27cm. The engineers were aware of this, but made a sign error in the calculations, so the difference was doubled to 54cm rather than being cancelled out.

Normally with big infrastructure projects, if something is built in slightly the wrong place, you end up having to live with it. It is just too expensive and disruptive to take it down and start again. Had the *whole* bridge been built to the wrong height, the error would have been quietly buried and the good citizens of the Laufenburgs would probably never have noticed. Building *half* the bridge to the wrong height was a wee bit conspicuous.

So the project became famous as the bridge where the two halves didn't match. But that's not why it should be famous.

When the error was discovered, the engineers got together and sorted out a fix, which apparently involved lowering the arm being built out from the German side. And they did it without having a row. Nobody played blame games. Nobody sued anyone. The budget didn't get blown and neither did the schedule.

Now THAT is worth being famous for.

The board controls the company. Or does it?

There is good governance. And then there are attempts at good governance that bear a close resemblance to hitting your head against a brick wall.

In mid-1994 the government of British Columbia announced the construction of three fast car ferries for the British Columbia Ferry Corporation.

BC Ferries decided to manage the project through a specially-formed subsidiary, Catamaran Ferries International Inc, incorporated in March 1996. A board of directors was appointed in May, including three independent directors with significant shipbuilding experience.

The CFI board lasted 10 months.

A review by the Auditor-General found that over those 10 months, the board:

- asked repeatedly for a full budget;
- stressed on several occasions the need for a contract between CFI and BC Ferries;
- asked to receive regular reporting against an approved budget, showing both forecast costs to complete and variances;
- asked for a construction schedule;
- pointed out that the original budget was for a different type of ferry, and questioned both that budget and the latest changes to it;
- stressed the need for a risk analysis of the current scope of the project; and
- noted that forecast costs continued to rise, and that the scope of the program had increased without an increase in budget.

Despite their efforts, the quality and quantity of information received by the board did not improve.

The board sought advice from BC Ferries on whether it was entitled to sack its chief executive (who was employed by BC Ferries rather than CFI).

They were told no.

In March 1997, the chief executive promised the board a complete budget by the time of their April board meeting.

Didn't happen.

Instead, the board members were invited to resign, which they duly did.

The CFI chief executive became a member of the newly-reconstituted board. The three experienced independent directors didn't.

Not the finest hour for project governance.

The three ferries? Three years behind schedule, at a cost of about $460 million, more than double the original $210 million estimate. The last one was built but never commissioned, because by then the government had given up on them. The fuel consumption was high, loading and unloading took too long, they kept breaking down and the passengers didn't like the accommodation. The ferries were auctioned off in 2003 for the somewhat pathetic total of $19.4 million.

Does a PPP do what it says on the tin?

A key benefit of a PPP is supposed to be that, with a fixed price for, typically, five years of construction followed by 25 years of operation and maintenance, it incentivises the contractor to optimise whole-of-life cost.

No more contractors building a shoddy asset because they know someone else will have to pick up the higher maintenance costs. No more contractors walking away on completion and ignoring the needs of the operator.

With the contractor bearing the long-term responsibility for operations and maintenance, it's in their interests to ensure that whole-of-life costs are optimised, not just construction costs.

The theory is good. Does it work in practice?

A PPP requires the contractor to finance construction. The borrowings are repaid with the availability payments the contractor receives during the operating period. This means the period of maximum drawdown and maximum interest payments is the period immediately prior to construction completion.

If anything has gone seriously wrong – and sometimes it does – the contractor may be staring ruin in the face if they can't get construction completed quickly enough to start operations and open the money tap.

In these circumstances, a PPP places an enormous temptation on the contractor to cut corners, permit defects to go unremediated and generally shunt problems off to the operational phase.

The whole-of-life cost motivation hasn't gone away. The contract still says what it says.

But as the interest payments keep ratcheting up, the contractor is balancing the prospect of going bust right now against the possibility of going bust some years down the track. Whole-of-life cost? Let's worry about that later.

What is it about confidential information?

Thirty-odd years ago a partner in a UK law firm blithely told me how he had fortuitously secured an advantage in a negotiation: he happened to be in a hotel standing beside the fax machine (remember them?) when a fax came through for the other side. He read it and made use of the information.

A partner should have known better than to breach professional ethics like that, let alone boast about it.

But it happens.

The inquiry into the Queensland Health payroll disaster back in 2013 recorded an incident during the procurement process where information about the various bidder proposals was mistakenly placed on an open drive to which bidders had access. One of the bidders heard about the error and went looking for the material. He didn't find it, because the error had been discovered and the material removed to a secure drive. But he saw no issue with trying to find it, and openly sent a somewhat plaintive email to his boss describing his efforts, saying "... *looks like we were just a little bit too late.*"

People who would never lie or steal or even jaywalk will nonetheless gobble confidential information like sharks in a feeding frenzy. How come?

Run down by the disputes truck

There's nothing wrong with risk avoidance as a strategy. By all means take the overbridge instead of trying to run across a six-lane highway at truck level. I totally approve.

But if your strategy is risk avoidance, you have to implement the strategy before the risk has arrived. After you've been hit by the truck is too late.

Dispute Avoidance Boards are increasingly used in megaprojects. Their purpose is – surprise! – to avoid disputes. The whole idea is to spot the issues that are at risk of turning into disputes, and help the parties to resolve them early, before positions become entrenched and costs have escalated into the stratosphere. The costs of arbitration on megaprojects can be horrendous because the issues are usually complex, the stakes are high and the cases can take years to resolve.

You don't appoint an arbitrator until a dispute has become intractable. But a Dispute Avoidance Board is there to stop the intractable dispute from arising in the first place. If you wait to appoint one until you have a dispute, you've lost the plot.

And if you appoint a Dispute Avoidance Board but then hide the issues from them or fail to meet with them until the truck is bearing down on you, they're not very effective.

Yes, there's a cost to building an overbridge. That doesn't mean it's not a good strategy.

Getting hit by a truck can cost much, much more.

Financing England's canals

We complain about banks, but they do come in handy for financing infrastructure, and (given a half-decent project) these days they will even queue up for the opportunity. It wasn't always that easy.

Back in 1762 Francis Egerton, 3rd Duke of Bridgewater, managed to obtain an Act of Parliament giving him the right to build a canal from his coal mines at Worsley in Lancashire to the factories of Manchester. It didn't give him the money.

Since canals were a new thing in Britain, the Duke struggled to fund it. Half his critics thought it was a foolish venture that would never pay, and the other half was convinced the canal was an evil idea that would ruin the land-based carrier trade and drive innkeepers bankrupt. The banks didn't want to know.

Things got so tight that on several occasions in the course of construction, the Duke had to send his agent to go round to his tenants and collect whatever amounts of rent he could so it could be paid out immediately to settle the weekly wages of the workmen.

This first canal, finished in 1761, was a huge financial success, and on the strength of its revenues Child's Bank was finally persuaded to fund an extension to Liverpool. (Traditionally, banks prefer lending to people who don't need the money.)

At last able to operate with Other People's Money, the Duke carried on canal building. By the time of his death in 1803 he was said to have become the richest nobleman in England.

For Georgette Heyer fans, the 3rd Duke of Bridgewater was not one of the two Dukes married by the elder of the famous Gunning sisters. He did become engaged to her in 1758, but for some reason the engagement was broken off later that year and he retired to his estates to build canals.

What is governance?

Anything which is good for a project runs the risk of being labelled "governance".

I have even seen "efficiency" described as a principle of governance, for instance, which strikes me as bizarre.

Not that efficiency isn't a good thing, but a principle of governance? No. It just demonstrates "governance" has become an over-used word that means what people want it to mean.

In the context of setting up a megaproject, my own take on governance, although perhaps idiosyncratic, is very simple.

At some point, a CEO or Minister will take up a pen to sign a contract that will commit the government to billions in expenditure and to public vilification if it all goes wrong. And the signatory will always ask, often very nervously: "Is it really okay for me to sign this?"

Governance exists to provide the answer to that question.

Has the project team done everything that should have been done? Have all the bases been covered? Has the Minister's backside been covered? Governance is how you know.

So, management and governance are not the same thing.

Management is doing stuff. Governance is checking that the right stuff is being done.

When is a fixed price not a fixed price?

Ask the City of Edinburgh Council (CEC). Or possibly don't. After the publication in September 2023 of the Edinburgh Tram Inquiry report, their response is likely to be explosive.

"With these final fixed price contracts now completed all parties can now proceed to delivering this project safely to programme and budget." Those were the words of the media release issued on 14 May 2008, after CEC had executed a contract for the construction of a £508 million tramline.

Media releases are unreliable sources of information. Yes, there was a headline price which was ostensibly fixed, but the construction contract was only "fixed price" in a manner of speaking. A highly inaccurate manner of speaking.

The contract included a schedule which set out the assumptions on which that price was given. Both **tie**, the CEC subsidiary which acted as CEC's procurement and delivery agent, and the contractor knew the assumptions included a number of statements that were not correct when the contract was presented for signature.

Setting out assumptions in a contract is a legal device, used where a situation is uncertain, to make it clear where the line is being drawn on which party takes the risk where an assumption turns out not to be correct. It is not customary to use them where the assumptions set out a position which is not merely uncertain but already known to be wildly and appallingly inaccurate.

The costs ended up exceeding the fixed price by more than 40%, even though the finished line was two kilometres short of the original destination.

No construction contract price is ever truly fixed, and for a megaproject the difference between the fixed price and the actual cost can be, well, mega.

CEC could tell you it pays to enquire into the circumstances in which the fixed price can be exceeded.

Be nice to losing bidders

Why?

Losing bidders are key to the success of the procurement process. They provide the competition that will deliver value for money.

The winning bidder doesn't make you a great offer because they like you, they make a great offer because they think if they don't, some other bidder will beat them to the contract. This generally gives you a much better deal than you could ever have managed in a 1-on-1 negotiation.

On a megaproject the losing bidder will spend millions on the bid process, push the winning bidder to give you the best possible deal, and will still walk away with nothing.

That deserves respect.

Give them a proper debrief.

And if respect isn't reason enough to convince you, consider it good self-defence. A bidder given believable reasons for coming second is much less likely to sue you than one that can't understand how they could possibly have lost.

Local industry preference

Last bastion of iconic local industry heroically wins bid against foreign competition.

Right, that procurement is doomed then.

As indeed it was. The procurement in question was the purchase of two ferries – huge, cargo-carrying vessels, not commuter transport – for delivery in 2018 to serve the Hebrides. The contract was awarded in 2014 against heavy foreign competition to Ferguson Marine, the sole survivor of the many shipyards that dominated Port Glasgow and Greenock when Scotland was the centre of the global shipbuilding industry.

Well, sort of survivor. It had gone into administration that year, with most of the shipbuilding workforce being made redundant by the administrators. A passing Scottish billionaire bought Ferguson Marine out of administration just in time to bid for the ferries contract. Lo and behold, Ferguson Marine was awarded the £97 million contract despite a clear recommendation to the contrary by the Government purchasing agency, Caledonian Maritime Assets Limited (CMAL).

Audit Scotland complained there was "no documented evidence to confirm why the Scottish Ministers were willing to accept the risks of awarding the contract". But no prizes for guessing it had something to do with being unwilling to be seen to kill off the last shipyard in Port Glasgow.

Killing it might have been a kindness, because the project was and still is a disaster. The government could have terminated the contract several times over, but when the ferries were already a year late decided instead to nationalise the shipyard. Performance remained poor. CMAL has spent more than four times the original contract price and so far received only one of the two ferries – the Glen Sannox, which finally entered service in January 2025.

It is not necessarily wrong to prop up a local industry with government contracts (although in some jurisdictions it is at least theoretically illegal) – but if you admit that is what you are doing and plan accordingly, you are more likely to get the desired outcome than if you try to pretend you are conducting a normal commercial transaction.

The sad demise of meaningful recitals

Forty years ago, as I was embarking on what I then expected would be a whole-life career as a lawyer (got that wrong) the front page of a contract was labelled "Recitals".

Recitals were a series of statements beginning with the word "Whereas", which set out what the contract was all about. Whereas [Party A] wants to procure a new building for its international headquarters, and Whereas [Party B] has agreed to design and construct said building or whatever.

My supervising partner used to say, "If you can draft the recitals, you can draft the contract," meaning that if you could say in simple language what the parties were trying to achieve, there was a reasonable chance the contract you drafted would deliver what the parties wanted.

I had occasion to look at a contract the other day and was sad to see that the Recitals, re-labelled "Background" and missing the whereases, consisted of four brief statements, at the end of which I knew that ServiceCo had agreed to deliver the Service Package.

I turned to the definitions section and discovered that the Service Package meant the performance of the Service Package Activities in accordance with the Service Package Documents. The Service Package Activities were defined as things ServiceCo was obliged to do under the Service Package Documents. The Service Package Documents were defined as a list of documents with helpful names like "this Deed" and "the Equity Documents".

Did the contract deliver what the parties wanted? I have no idea.

Short is not always sweet

Twitter hadn't been invented, there was no 140-character limit on communication, so I don't know why the memo went out with just 5 words, 30 characters.

Milo O. Frank, author of the book *How to Get Your Point Across in 30 Seconds or Less* thought it was terrific. I'm less impressed.

The memo was sent to all staff by an American businessman just before Thanksgiving. It said "Closed for the holiday. Enjoy!"

In a book about how to get a point across quickly, I suppose it must have seemed worthy of praise.

But the company was a large one and I wonder if his payroll team was quite so impressed. They would probably have been inundated with queries like, "Will I get paid?", "Does the day come off my vacation entitlement?" and "I've already booked it as annual leave, what do I do?".

Short is good. Sometimes it's not enough.

Empire State Building

You know what the really remarkable thing about this project was?

When the decision was made to construct the Empire State Building, the time given to complete the then tallest building in the world was "eighteen months from first sketch to last rivet", as Flyvbjerg and Gardner put it in their book, *How Big Things Get Done*.

The Empire State Building opened on time on 1 May 1931, 17% under budget. Impressive, but that wasn't the remarkable thing.

The remarkable thing was they didn't start work on the construction site until the design was finished. And I mean finished.

Flyvbjerg and Gardner quote a 1931 publication, *The Empire State*, in which the company boasted that: "the architects knew exactly how many beams and of what lengths, even how many rivets and bolts would be needed. They knew how many windows Empire State would have, how many blocks of lime-stone, and of what shapes and sizes, how many tons of aluminum and stainless steel, tons of cement, tons of mortar.

Even before it was begun, Empire State was finished entirely—on paper."

How many organisations today would have the discipline to hold off commencement of construction until the design was 100% finished?

I've seen government bodies decide on a strategy of letting the infrastructure design contract separately from the construction contract – and then charge ahead with work under the supposedly "construct-only" contract while the design is still a work in progress.

The usual excuse is to avoid delay, but guess what? The schedule is inevitably cruelled by the variations and re-work required. So is the budget.

Design first. Then build. How hard can it be?

Should lenders be accountable for safety?

Project financiers want projects delivered on budget so they can get their money back.

The ideal way to deliver to a budget is to get things right first time, on time, solve problems quickly and find inexpensive ways to deliver what is required.

That's the ideal. It doesn't always work that way. Financial pressure can instead drive people to conceal problems, cut corners and take unjustified risks.

Remember the Beaconsfield gold mine in Tasmania in 2006 where one miner died and two were trapped for several days before finally being rescued? Afterwards, some of the workers claimed that when they had raised concerns about safety corners being cut, the response was, in effect, "we have to do this because if we don't get out enough ore to pay the debt, the company goes under and we're all out of a job, so shut up and get on with it". I don't think those allegations were substantiated in court, but it illustrates the potential problem.

If you apply severe financial pressure to an organisation which doesn't have both adequate safety systems in place and managers of character to apply those systems, evil will be done.

Has anyone ever seriously tried to hold lenders responsible for project safety?

Perhaps they should.

Only one thing beats a photo opportunity

We all know creating photo opportunities for politicians is part of the project manager's day job. A hard hat, a hi-vis vest, and a photographer – what more do you need?

Something for the politician to do.

TV news doesn't want a photo, they want video. Social media has somewhat blurred the primacy of the national news, but it's still the case that a video will always trump a photo.

Turning the first sod, laying the first brick, anything that provides a bit of movement for the video cameras.

Even in the procurement phase you can always find something. I still have in my possession a green linen tablecloth which starred on Australian national television when the NSW premier announced the award of the Waratah rolling stock PPP: he pulled the cloth away with a grand flourish to reveal a scale model of the selected eight-car train.

If you really want to keep the politicians happy, don't just find them a photo opportunity, find them a video opportunity.

Works every time.

Working off the wrong baseline

The Edinburgh Tram Inquiry, published in 2023, found the designers of the tramline were engaged in September 2005, three months before the Scottish parliament had finished considering the project.

During those three months, significant changes were made to the original plans and sections. At Haymarket Yards, the route changed completely. There were changes to the horizontal and vertical limits of deviation.

Nobody told the designers, who continued to work from the baseline they had been given.

The preliminary design phase bled inefficiently into the detailed design phase, at considerable expense. The issues arising from the initial failure to inform the designers were more or less sorted out, only for the designers to discover that further specification changes had been agreed with the bidders without reference to the design team.

Infuriating.

And expensive. The Inquiry estimated that the delay and rework added 30% to the design costs.

PART 2

The wrong project governance framework

When the UK Department for Transport unsuccessfully attempted to let the InterCity West Coast franchise in 2012, the internal auditors produced a *"Rail Governance Map"*. This was issued about a week after the tenderers had submitted their bids – that is, when the tender process was in full swing, and good governance was essential.

Good work.

Except that the subsequent report from the National Audit Office found that, unbeknown to the internal auditors, the actual governance framework was quite different.

The actions of a *"Contract Award Committee"* ultimately caused the tender process to be cancelled, since it determined the amounts of the subordinated loan facility in a manner not in accordance with the guidance provided to bidders.

But the Committee's terms of reference said its role was *"providing assurance on the procurement process"*. Determining the subordinated loan facility amounts was hardly an assurance role.

It seems the committee members based their belief in their authority on the process charts issued with the guidance to bidders. But none of the subsequent inquiries managed to establish who took or approved the decision to issue the guidance in the first place.

The committee failed to meet at all for five months during the tender process. During those months the project was without a Senior Responsible Owner for an extended period (this unfortunate omission was eventually picked up by an external review) and in that time *"emerged"* the critical, unattributed, decision around using a financial model that was not fit for purpose.

A governance committee that is supposed to provide assurance but fails to meet and is mistaken in its remit is unlikely to produce much in the way of assurance, and so indeed it proved. The losing bidder sued and the tender process had to be cancelled.

There is no universal perfect governance framework for a megaproject. But if you don't know what yours is, it's the wrong one.

Hyperbolic discounting

The brain can do some pretty weird stuff when it comes to size perception.

When cameras first became widely available, decades before Photoshop was invented, a favourite trick was to position two people, one close to the camera, one further away up a hill, so it looked as if the distant person was a miniature human being standing on the outstretched hand of the closer one.

When it comes to dollars, distant ones are also perceived as smaller than closer ones. It's called hyperbolic discounting. Passing up an immediate benefit for a larger future benefit is hard to do: only compulsory contributions explain why there is so much money in Australian super funds.

People normally talk about hyperbolic discounting in relation to the perception of benefits, but it applies equally to costs.

In public sector terms, a cost incurred now looks bigger than a cost pushed off onto the next government.

A cost pushed off onto the next generation looks even smaller.

Hello climate change.

Evaluating references

An unsuccessful foreign bidder for a major NSW contract some years ago told me he had been completely bewildered by the approach taken to the evaluation.

One of the criteria had been about the ability to work with the client collaboratively. As the popularity of collaborative contracts increases, a criterion of this nature is becoming increasing common. (Although it should have been more popular years ago — the ability to work well with the client has always been an important success factor, no matter what the type of contract.)

At the debrief, the bidder was told they had scored poorly

because they hadn't provided much evidence of their collaborative skills.

The bidder thought there must be a language problem. They prided themselves on never having a dispute with a client. They had provided contact details for several of them, in absolute confidence the clients would provide rave reviews and as much evidence of collaborative skills as anyone could want. What had gone wrong?

It turned out that, although the evaluators had asked for references, they didn't use them to help select the preferred bidder. Instead, the references were used after the decision had been made, to confirm the preferred bidder had been telling the truth.

Which meant that what the bidder had regarded as the world's best evidence, the evaluators had regarded as evidence not worth considering.

I think I'm with the bidder on that one.

The core business of a railway is … ?

Possibly not making drawing pins.

I joined Railtrack, then the owner of the former British Rail network, in 1995, a year before it was privatised. The cork noticeboard in my London office still had brass drawing pins with "BR" embossed on the head, manufactured in-house somewhere oop north (wish I'd kept one as a souvenir).

British Rail had done pretty much everything it was possible to do on a railway, but the 90s restructuring tried to reduce that to the core business. Most of the freight, engineering and manufacturing activity was being or had already been sold off, leaving Railtrack to own and manage track and signalling, some track maintenance companies, a few companies to own and lease out the rolling stock, and lots of different train operators.

The post-BR operating model took a battering over the next couple of decades. Part of the problem was that it wasn't entirely clear whether the core railway business sat with the train operators, who were the ones providing the services to passengers, or with Railtrack, which managed the timetable.

It is perhaps not possible to identify whether the various failures over that time were structural or just down to rubbish execution, but an overall trend emerged: with each failure, bits which had been carved off as "non-core" were gradually sucked back into the centre.

There is now a plan to bring all passenger operations, train and track back into an integrated railway to be called Great British Railways. Perhaps it's all core?

But I don't think anyone is proposing GBR will manufacture its own drawing pins.

Professional integrity is not an absolute

Most people think of professional integrity as something you either have or you don't.

I don't think that's right. I also don't think it's helpful.

There are two elements of professional integrity:

1. having high professional standards;
2. being prepared to uphold those standards in the face of considerable personal cost, such as the loss of a client.

When we admire someone for their professional integrity, we are effectively saying that either their professional standards are higher than ours, or their pain threshold is higher. Or both.

Neither of these are absolutes. You can raise your standards through education, for instance, and raise your resilience to cost by improving your financial position.

If you recognise professional integrity is not absolute, you absolutely can improve it over time and you can help others to improve too.

Go for it.

Safety culture

At Clapham Junction in 1988, one train ran into the back of another and derailed. A third, travelling in the opposite direction on the adjacent track, hit the derailed train. Thirty-five people died.

The proximate cause was an old wire in a signal box. It had been disconnected but not secured: it slipped and re-connected, turning a red signal green.

The proximate cause is never the whole story. The report of Anthony Hidden QC, the Chairman of the subsequent Inquiry, made 93 recommendations covering a range of activities from fatigue management to accident response.

The technician who had disconnected the wire was distraught. The Inquiry was advised that good practice would have been to cut the disconnected wire short so it could never reach its old position, and to tie it back so it couldn't move at all. The technician had done neither of those things. But then he never did. He just followed his usual practice of pushing the old wire out of the way. Staggeringly, he had been doing his job for 16 years, and never once in all that time had anyone ever told him that what he was doing was wrong or unsafe, or even noticed he was doing it wrong. Checks that would have prevented the accident were not carried out.

The managers of British Rail, who probably thought of themselves as good, hard-working people, were forced to confront the realisation that the deaths of those 35 people were essentially their fault. As Hidden put it:

"The best of intentions regarding safe working practices was permitted to go hand in hand with the worst of inaction in ensuring that such practices were put into effect. ... It has to be said that a concern for safety which is sincerely held and repeatedly expressed but, nevertheless, is not carried through into action, is as much protection from danger as no concern at all."

Culture is not about what you say, it's about what you do.

Payroll problems

What did the failed implementation of the Queensland Health payroll system and the ACT HRIMS payroll system have in common?

The pre-existing pay arrangements were a complete dog's breakfast.

A report by KPMG noted that the 85,000 staff of Queensland Health were employed under two different pieces of legislation, covered by 12 industrial awards and impacted by 6 different industrial agreements. The effect was to create over 200 allowances and up to 24,000 combinations of pay. To make things worse, several thousand employees had concurrent employment arrangements – they held more than one position, each with different terms and conditions.

Queensland Health did manage to launch their new system in 2010, but probably wished they hadn't. After the launch, thousands of staff received the wrong amount, or nothing at all. It took months to develop a functioning payroll system, with more than 400 additional payroll staff engaged to deliver the estimated 200,000 manual processes required to deal with an average of 92,000 forms every fortnight. This for a project that was supposed to result in staff cost savings.

The Commission of Inquiry described it, generously, as a *"catastrophic failure"*.

Afterwards, the Auditor-General of Queensland recommended that agencies wishing to replace their payroll system in future should simplify award structures prior to implementation, to remove complexities which would impact on the effectiveness and efficiency of the payroll process.

The ACT obviously failed to get the memo. The starting point for HRIMS was, if anything, worse.

A report from the ACT Auditor-General found there were eighteen separate Enterprise Agreements covering diverse workforces, occupations and employment conditions across the ACT Public Service. An additional 13 Remuneration Determinations covered unique payroll terms for executive staff, Public Office holders and board members. There were more than 11,000 leave processing rules across the Territory for 54 distinct types of leave and 21 classes of employee, at an average of 524 rules per employee class. Et cetera.

Work was stopped on the HRIMS Program in June 2023. At least

$77.63 million had been spent and only one module, the Learning Management System, was able to be delivered.

Sigh.

Audit reports often contain useful recommendations. Be nice if more people paid attention to them.

New river, old approach to planning

NIMBY-ism was alive and well at the beginning of the seventeenth century.

London desperately needed a supply of clean water, and it was proposed that an artificial waterway be constructed from the Chadwell and Amwell springs near Ware. The City of London, whose residents were fed up with hauling water in buckets, petitioned Parliament to support the project.

Landowners on the route were appalled. They also petitioned Parliament, claiming "that their meadows would be turned into bogs and quagmires and arable land become squalid ground; that their farms would be mangled and their fields cut up into quillets and small pieces; that the cut, which was no better than a deep ditch dangerous to men and cattle, would upon sodden raines inundate the adjoining meadows and pastures, to the utter ruin of many poor men".

Parliament ignored the landowners' petition and passed a series of acts conferring the water rights on Hugh Myddelton, a prominent goldsmith and the Royal Jeweller (he was also an MP and no doubt voted in favour of giving himself the rights). Alas, the NIMBY landowners continued to make trouble and Myddelton was soon in financial difficulties.

King James came to the rescue with funding for the project. Although, kings being kings, not until Myddelton agreed to give him a half share in the profits.

James owned Theobalds House, which was on the proposed route, and did not object to its extensive parklands being traversed by the new waterway. The other landowners promptly fell into line. It is not clear whether they simply wished to please the king, or whether they were terrified of being asked to please the king by giving him their land, which was how James had acquired Theobalds House in the first place.

The only evidence the seventeenth century NIMBYs were correct in thinking the new waterway would be dangerous to men and cattle was provided on 9 January 1622, when a group of revellers went to look at the ice on the New River. One of them fell in head first, so that his companions could only see his boots. Fittingly, the reveller in question was King James, probably drunk as a skunk.

The waterway was completed in 1613. It took twenty years to pay its first dividend (too late for James, who survived his dip in the river but died in 1625) but became extremely profitable. It is still, four centuries later, known as the New River. Now run by Thames Water, it currently provides about 8% of London's water supply.

Good thing the NIMBYs lost.

Non-standard equals sub-standard

Why are governments always so reluctant to accept a standard design for anything?

Let's face it, when it comes to products like trains and trams, suppliers may have more experience and expertise than government, probably across many different countries, and may have a much better feel for what the end users would really like.

When you request modifications of a standard form product, you add cost. If the standard form tram comes with CCTV, for instance, it will be more expensive to buy the tram without CCTV. This is counter-intuitive but true.

A CCTV-free tram might have been cheaper if the tram had been designed without CCTV originally, but that is not the situation you are in. If you change the design, you not only have the cost of the re-design, the manufacturer has to stop the production line to re-program the machines and stop it again to put the settings back to standard when your job is over.

Plus, the fact the manufacturer decided to put CCTV into the standard model means most of its customers want it. Why should your customers be different?

One tram manufacturer politely, for a fee, modified its standard design for a public sector customer which demanded the removal of poles from the area near the doors.

The company simultaneously prepared a quotation for installing the poles as a modification, knowing perfectly well that tram passengers are more concerned with being able to hang on to something than with the aesthetics of the vestibule.

It didn't have to wait very long before the quotation was requested. No surprises there.

Non-standard? Or sub-standard? Sometimes they're the same thing.

What isn't keeping you awake at night?

I'm always looking for good questions to ask in project reviews. How can you make the interviewee really think about the answer, rather than just trot out the standard response?

"What keeps you awake at night?" is often quite good for getting people to open up about the things that are really bothering them.

It is, however, completely useless for the problems a review is most interested in discovering. The issues that people should be worrying about – but aren't.

Back in the early 1980s, British Rail had a serious early morning derailment at a set of points south of Victoria station. One of my former colleagues had to deal with it.

He called up the engineer responsible for that section of track and told him what had happened, without giving details of the location. "Oh God," was the response. "Not..." and he named the set of points.

That engineer knew. He knew the state of the points was a disaster waiting to happen. But he hadn't taken steps to prevent it. It hadn't kept him awake at night.

So here's a different question. Imagine you get a panic call from your boss at 4 a.m. Which potential disaster springs to mind?

Maybe that's the one you should be doing something about.

If you can't fix it, ring-fence it

If you're still trying to sort out the specification, you're not ready to go out with a call for tenders. Sometimes you have to do it anyway.

On one New South Wales contract, there was an issue about which of two standards the contractor would be required to comply with in relation to one aspect of the works. Resolving the issue could only be achieved by undertaking an initial study.

Ideally, the study would have been carried out by the agency before any request for tenders was issued, but the amount of time involved was incompatible with a politically acceptable procurement timetable. (Sound familiar?)

Requesting prices for an uncertain specification would be asking for trouble. Bid prices are meaningless if you don't set out the contract terms in full. How to square the circle?

The problem was dealt with as follows:

- The contract provided for the study to be carried out by the contractor.
- The bidders were asked to price for the works on the assumption the less onerous standard would be required.
- The contract included a specific variation provision that would be triggered if additional works were required under the more onerous standard.

Not ideal, because if the variation procedure had been triggered it would have cruelled the schedule. But it did at least allow the bidders to come up with a firm price.

In-house expert v. External expert

The difference between an in-house expert and an external one? That is, apart from the deeply irritating truth that the Board is likely to be more impressed with a report prepared by externals, regardless of actual quality.

The difference is that people forget internal experts need remits too.

When you engage an external consultant, the procurement process requires you to give the matter some thought. What is it you want them to quote for? What is the deliverable? How long will it take to produce it? What does their professional indemnity insurance need to be on the hook for?

Internal expert? Maybe you drop by their office and say, "Can you take a look at this for me?" And it just grows from there. Stuff gets produced, but never with clear deliverables or accountability. And often without any understanding of the context.

The Laidlaw report identified this as an issue in relation to the failure of the procurement process for the InterCity West Coast franchise in 2012, where evaluation of the bids was compromised by the inappropriate use of a financial model (known as the GDP Resilience Model) which was not fit for purpose.

Laidlaw wrote:

"Evidence *from interviews suggests a culture whereby each team tended only to consider a narrow set of issues relevant to its functional area without a clear view on the implications for the overall project. As an example, while individuals from the rail analysis sub-team developed the GDP Resilience Model, they appear to have believed they were only responsible for providing modelling input and were not responsible for sense checking the outputs."*

Internal experts may be just as expert as external ones.

But if you expect them to operate without a clear remit you won't be getting their best work.

Selecting a preferred bidder

Megaproject contracts are complicated. However fair and reasonable and standard the terms (and you should always do as much as you can to ensure the terms are fair and reasonable and standard) there will always be something you can't settle without negotiating with the bidder.

Which can be a problem when it comes to selecting a preferred bidder.

Consider the scenario in *Figure 1*. Bidder 1 is clearly ahead of Bidder 2. Bidder 2 potentially offers better value for money, but there's no guarantee they would be ahead of where Bidder 1 is now when the contract is finalised. If you select Bidder 1, you're potentially missing out on a lot of value. If you select Bidder 2, you risk the process being challenged by Bidder 1, who is in front at the point of decision.

Which should you select?

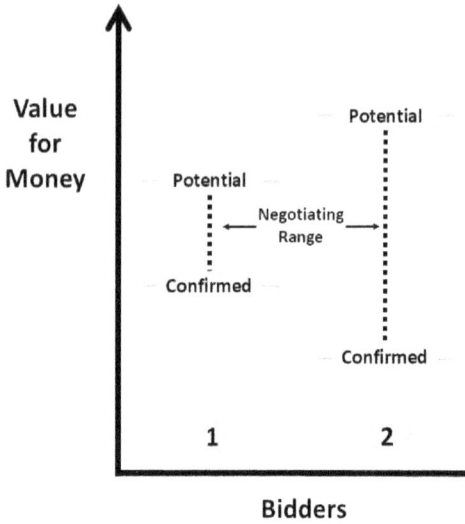

Figure 1: Evaluation scenario

Once a preferred bidder has been announced, competition is effectively gone. Even if the runner-up is still in the wings and theoretically able to step in if negotiations with the preferred bidder are unsuccessful, this doesn't create much competitive tension. What little there is will dissipate quickly, since the practicality of the alternative diminishes as time passes. Your negotiation range shrinks with it.

The best answer in these circumstances is probably not to select either bidder – yet.

When the decision is close, you won't be certain you're making the correct selection until you've reached the point where you have a contract you can sign. Yes, getting to that point with two bidders can seem daunting. Finalising two sets of megaproject documentation is no small task.

On the other hand, my experience is that negotiating with two bidders generally takes less time than negotiating with one. Once appointed, a preferred bidder will feel secure enough to seek more concessions from you, and negotiations can drag on and on. Knowing a competitor is still in the race discourages a bidder from raising all but the genuinely important issues.

Definitely an option to consider.

Attention to detail

I had occasion to travel into Sydney by train recently and noticed Sydney Trains was running a pram safety campaign. Ads and announcements warned passengers to park prams parallel to the platform with the brakes on.

It reminded me of a similar campaign run by RailCorp, predecessor of Sydney Trains, a dozen years ago. Tickets were still printed manually in station ticket offices, and there was a space on each ticket where a public service message could be printed. This space had been commandeered for the pram safety campaign.

The message was short and to the point:

TRAVELLING WITH A PRAM? TAKE CARE.

Alas, someone had miscounted the number of characters available. When the message appeared on the tickets, it read:

TRAVELLING WITH A PRAM? TAKE CAR

Oops.

Breaking the whisky bottle

Announcing completion of a project while there is still work to do is pretty much universal. If you've done enough to commence operations, you can quietly ignore the snagging list, particularly if you're a Minister and desperate to get an announcement in before the next election. But it doesn't always work.

The classic example of over-enthusiastic PR surrounded the Glen Sannox, the first of the two ferries to be delivered in Scotland by Ferguson Marine (see Local industry preference above). The ferry, scheduled to be delivered in May 2018, finally entered service in January 2025.

The over-enthusiastic PR surrounded the initial "launch" in 2017, when it was floated out of the dry dock for works to be completed elsewhere. The occasion was one of some celebration and nobody noticed anything wrong at the time. But two years later, the Scottish press had an absolute field day when it was discovered the bridge of the still-undelivered Glen Sannox had had no windows ("Nothing to See Here" was the headline in *The Scottish Sun*.)

Some PR enthusiast had just slapped on a bit of black paint to make the Glen Sannox look more finished that it really was. The non-delivery of the ferry was quite embarrassing enough for the Scottish government, but the discovery that the windows were only painted on made them a complete laughingstock.

So much so that when the second ferry finally left dry dock in April 2024, with windows, neither First Minister Nicola Sturgeon nor Transport Secretary Humza Yousaf was anywhere to be seen. The honour of smashing the whisky bottle on the hull (not champagne, this is Scotland we're talking about) was instead given to Beth Atkinson, one of the yard workers, who had just qualified as a welder. Congratulations, Beth. May you have a long and successful career.

Assume a can opener

That's a joke about economists, but also an approach to problem-solving seen more often than most people admit. Here's a classic example.

The Inquiry into the much-delayed Edinburgh Tram project found the schedulers had a solution to delayed utility relocations that would have made an economist proud.

Each time they had to reissue a revised schedule, they adjusted the productivity assumptions. Not, as you might expect, to reflect any level of productivity actually achieved on the project. No, they decided instead to reflect a level of productivity that would deliver the desired end date.

Or, in economist-speak, they assumed a can opener.

Guess what. It didn't materialise.

Should you ever find yourself on a desert island with a can and no can opener, I'm told you just keep rubbing the base of the can against a flat rock and it will eventually come off. But however many times you put unrealistic assumptions into your project plans, they don't get any more realistic.

A fixed formula for megaprojects?

Many agencies have a standard form evaluation spreadsheet and mandate its use. Before you evaluate the bids, you decide on the preferred weighting between technical and pricing scores for this particular procurement, and that weighting is locked into the spreadsheet.

Once the bids have been scored by the evaluators, technical scores for each tenderer are entered into one column, the pricing into another column, and the spreadsheet tells you which bidder should get the contract.

It works quite well for small procurements. It limits the opportunity for biased assessment and gives peace of mind to the evaluators. No worries about whether you have the right answer. The spreadsheet has spoken.

Does it work for megaprojects?

You do need some way of integrating scores for price and technical criteria. Does higher cost outweigh lower quality, and if so by how much? The problem is that these standard form evaluation spreadsheets assume you are using the agency's standard contract.

Guess what. You aren't.

Bidders will mark up the contract (which is in any event unlikely to be the agency standard form) in all sorts of different ways, potentially having a significant impact on value for money.

Some of those changes may be readily quantifiable in dollar terms. But often the clauses amended are ones that will never be triggered except in a situation so dire neither you nor the bidder would sign the contract at all if either of you thought there was any real chance of it coming to pass.

Try putting that in spreadsheet.

Actually, I suggest you don't. Megaproject bids are capable of throwing up more complications than anyone could ever capture in a fixed formula.

So skip the fixed formula and give the evaluation committee a bit of discretion to exercise judgment in making its recommendations. Not enough to open the door to bias and manipulation. But enough to be able to deal with the unexpected.

It's a megaproject. Unexpected is what you get.

Design now or pay later

Disabled access, whether to buildings or trains, is one of those things that adds very little expense if you design it in up front. And costs a fortune if you don't.

Queensland ordered 75 six-car trains in 2014. The 2018 Commission of Inquiry found the project team had no adequate understanding of the requirements of disabled access legislation and didn't engage expert help to make up the deficiency. Early and genuine consultation with the disability sector might have put them on the right track, but that didn't happen either.

The biggest issue was that the specification issued with the request for tenders required only one accessible toilet on each train instead of two, which made compliance with the legislative requirements impossible. The one toilet cubicle that was installed did not meet the minimum size requirement.

The trains had hardly entered service when they had to be taken out again for modifications to the toilet cubicles: installing a second toilet, enlarging the original cubicle, adding help points and, in Wikipedia's somewhat unfortunate phrasing, installing "lights that warn in the event of an evacuation."

The cost, some $335 million, would have been a tiny fraction of that if the specification had got it right in the first place.

Levelling the playing field

When I led the Waratah train procurement, back in 2004, the NSW government decided to include construction of a new maintenance facility in the contract.

Train manufacturers generally have no construction expertise. Inclusion of a maintenance facility made it inevitable the successful bidder would be a consortium of some kind, with one member having the expertise in train manufacturing and one having the expertise in building maintenance facilities. Sounded like pretty standard stuff.

Problem. While the market for train manufacturers was international, the market for facility construction was local and there were more potential train suppliers than facility builders. Any train supplier who couldn't secure a dance partner would be handicapped out of the competition. The shortlist of bidders would effectively be determined by the facility builders' choice of which train supplier to partner with.

Not acceptable.

Instead, we made it possible for a bidder to be shortlisted without having construction capability.

A bit bizarre, given construction of the maintenance facility was a significant part of the $3.6 billion contract. But we took the view that the trains should have priority in the evaluation and that any train supplier making it to the shortlist would be able to attract a competent facility builder for the next phase.

It worked.

Recycling is good. Except when it isn't.

It is possible the oil company drilling in the Gulf of Mexico had genuine reason to believe their activities south of Florida represented a danger to polar bear habitats.

It is perhaps more probable that their risk register was a straight cut and paste job recycled from operations in Alaska. Not a good look.

A risk register isn't a box that you tick, it's a tool for managing the project:

- Creating it makes you think about what can go wrong and how to make things go right instead.
- Refreshing it periodically makes you think about what has changed and whether you need to start doing something different.
- Plus, it acts as a checklist for regular assurance that the planned risk management is actually happening.

Cutting and pasting doesn't add value. Not even to polar bears.

PART 3

Goodbye, megaprojects; hello, gigaprojects

We've been here before.

Word processors speed up typing. Contracts become longer.

Faster trains reduce commute times. People move further out.

Seat belts make accidents at speed more survivable. People drive faster.

Inevitably, things intended to make life simpler or safer are used to enable a greater degree of complication or risk.

So. We created the financing, engineering and contractual tools that enable us to do megaprojects. As the tools improve, will we just do megaprojects better, or will we use the tools to take on ever greater levels of complication and risk?

Um.

Gigaprojects it is, then.

Russian roulette

When you speak of "winning a contract", that's the language of playing games.

In some ways it IS a game. There are rules. You can get points on the evaluation scoreboard. You can give it your best shot and still end up with nothing when the whistle blows. Well played, better luck next time.

It's okay to think of the bid process as a game. But if you're the one setting the rules, you shouldn't be asking the bidders to play Russian roulette.

It's easy enough to shrug off a loss with a "win some, lose some" attitude if you know there will be another opportunity to win. The profits from one contract sustain you through the loss on another.

That doesn't apply if the loss is big enough to be a company-killer. Russian roulette is a dumb game to play, because you can only lose once.

One of the issues with megaprojects is that they come with mega-risks. Which means you can run into trouble with the general principle that a risk should be allocated to the party best placed to manage it.

Sometimes the party best placed to manage a risk is not also the best placed to survive it.

Expertise does not always come with the sort of balance sheet that can shrug off a loss measured in hundreds of millions. And sometimes disaster happens no matter how much expertise you bring to bear trying to prevent it.

Requiring a contractor to take on un-capped or un-shared risks too big for the balance sheet leads to bad practices, such as the contractor pushing risk further down the chain to subcontractors even less able to bear it.

You get best performance out of contractors when they have skin in the game. Just make sure it isn't the little circle of skin over the temple.

The sex life of badgers

It is probably a little embarrassing to have a contract structure determined by the sex life of badgers.

The project was Stage 1 of the Channel Tunnel Rail Link, now called HS1, which built the line from the Channel Tunnel to the edge of London.

The project contract was a monstrous thing: the documents, printed double-sided, eventually filled twenty-two 3-inch ring binders. There had been a PPP to build the line all the way to St Pancras (about 5 ring-binders' worth), but the contractor had run into financial difficulties and the project was completely restructured. The original PPP contract was preserved, but wrapped up in new contracts which required Railtrack, then in the unlikely role of white knight, to build stage 1 and take an option to build stage 2.

And then there were the badgers, who had most inconveniently dug their setts in the direct line of the new railway.

The British attitude to animals being what it is, the homes of people can be compulsorily purchased and demolished at any time of year, but badgers get a free pass during the mating season.

Presumably this considerate provision was omitted from the compulsory purchase legislation applicable to humans on the basis that, since the human mating season is year round, it would defeat the purpose of the Act.

Anyway, negotiations on the monster contract were prolonged perilously close to the start of the badger mating season. Failure to sign up before it began would delay the project by months, while construction waited upon the *amours* of the badger population. Not acceptable. But there's a limit to how fast you can draft 22 ring-binders' worth of contract documentation.

At the last minute, an extra contract was created especially for the removal of the badger setts, part of the monster contract suite but signed up in advance. Under it, the contractor started work immediately. The removal works would be paid for under the main contract, but the contractor would be able to claim separately if the negotiations fell through.

Thanks to the badger removal contract, the whole of Stage 1 was able to be delivered on time and on budget.

Not embarrassing at all, really.

Media massage

Some attempts to massage media response to projects go horribly wrong, like the launch of the Glen Sannox in Scotland with the windows only painted on (see Breaking the whisky bottle above).

Sometimes you almost get away with it. The Glen Sannox fenestration deficiency wasn't spotted at the time – it was a couple of years later that the media found out and had a field day bashing the project.

Sometimes you really do get away with it. Back in 2009 the Waratah prototype train for Sydney arrived at Newcastle docks to great press fanfare. I had left the project by then and just saw it on the news like everyone else.

It was years later before I learned the jolly banner slung on the train had been strategically positioned to hide an unsightly deformation of the steel shell, caused by a problem with the jigs. Its sole purpose was to conceal the defect from a press corps who would otherwise have turned a good news story into a project-bashing feeding frenzy. It worked.

Win some, lose some.

Dark revenue

Contractors don't care about client revenue, they only care about their own costs. Normally this doesn't matter.

The contractor is desperate to hand over the asset so they can stop incurring costs and the client is desperate to get the asset handed over so they can start generating revenue. When the infrastructure being constructed is new, the incentives are aligned.

For a replacement/renewal project, it's a bit different. There is inevitably a period when the client's revenue goes dark. The old asset has to be shut down, so it is no longer generating revenue, but the new asset is not yet operational, so it isn't generating revenue either. The period can be short or long, days or years, depending on the nature of the asset and the construction methodology.

The dark revenue period has two consequences. Other things being equal:

1. A construction methodology that reduces the dark revenue period is more valuable to the client than a construction methodology that extends it.
2. A risk mitigation measure that reduces the risk of delay during the dark revenue period is more valuable to the client than a risk mitigation measure that reduces the risk of delay while the original asset is still operating.

This rarely occurs to the bidders.

Their concern is getting the job done and construction costs down. On a brownfield railway project, for instance, they would much prefer a three-month shutdown to a year of weekend possessions, regardless of the impact on services and passenger revenue.

Should you force them to pay attention by making minimisation of the dark revenue period a criterion in the evaluation?

Maybe. But maybe not. Sadly, while the potential for revenue improvement is limited, the potential for cost overruns is almost infinite. Reducing construction delays – even ones aimed at preserving revenue for as long as possible – has a big impact on risk reduction. No comfort in generating a few more million in revenue if doing so provides a window for much more costly risks to materialise.

Buying time

Time pressure is a classic project killer. When urgency is trumping every other consideration, you know disaster is not very far away.

A dozen or so years ago, RailCorp had to re-let a contract for rolling stock maintenance in Sydney, a huge contract worth many millions. The expiry of the current contract was approaching, and while there was theoretically time enough to go through a procurement process, the time pressure would be intense.

So before we started the procurement process, we negotiated options for a couple of one year extensions with the existing contractor. There was still time pressure. The contract extensions would not be on as good terms as we could expect to get in the competitive market, so we wanted to move to the new contract as soon as possible. But we were no longer in a position where we absolutely had to sign a new contract immediately, and that made all the difference.

At around the same time, the UK Department for Transport attempted to re-franchise InterCity West Coast rail passenger services. The invitation to tender was issued in January for a franchise due to expire in December. Perfectly possible, but with very little margin for error.

Errors were made. The losing bidder brought an action for judicial review and the tender process had to be cancelled, with bidders taking home around £40 million in compensation. The incumbent got a hastily-negotiated two year extension, no doubt on favourable terms because Transport had nowhere else to go.

The subsequent Inquiry report was in no doubt of the key issue: *"the quality and robustness of the ICWC procurement was subordinated to an overriding pressure to complete the procurement on time"*.

There may have been no opportunity to negotiate extension options in advance. But the extra time it would have bought would definitely have come in handy.

Cost estimates way below outturn. Why?

Back in the 90s, managers at Railtrack (Network Rail's predecessor) reviewed inherited British Rail projects to discover why the cost estimates were so far removed from reality.

Deliberate "strategic misrepresentation" (low-balling costs to secure approval) wasn't the issue on this occasion – managers wanted to know at a fairly gritty level where the estimates were getting things wrong.

Curiously, it turned out British Rail had really been rather good at estimating costs. What they were terrible at was estimating scope. The projects cost more because the estimates left out chunks of activity.

I suspect it was a combination of a silo organisation and a poor investment process.

The silo organisation meant engineers would produce estimates that just said "plus property costs", rather than venturing over to the property department to find someone who could work out the likely property costs.

The poor investment process allowed projects to go forward when the estimates still said "plus property costs".

If you don't understand the scope, you can't possibly understand the cost.

Pockling the spreadsheet

London is the spiritual home of cowboy tradies, and actual home to a depressing number of them.

Back when I was living there, I was once presented with an invoice which didn't add up. Literally. The total at the bottom of the column was significantly more than the sum of the numbers in the column above it.

I can do arithmetic. And it is not totally unusual to find a tradie who can't. What threw me completely was that the invoice wasn't a text document, it was a spreadsheet. The Excel autosum function couldn't add up? Apocalyptic.

Fortunately, just before my brain went into total meltdown, I spotted that the row numbering on the left-hand side was not entirely sequential. Revealing the three hidden rows also revealed three more numbers in the column of figures, neatly bridging the gap between my arithmetic and the Excel autosum.

Needless to say, I paid the lower figure.

You kind of expect stuff like this from cowboy tradies. It is more disconcerting to encounter it in a megaproject.

In the Edinburgh Tram project, completed in 2014, the analyst doing the quantitative risk assessment to support the funding request was instructed by the agency to amend the spreadsheet – not to change the assumptions, but to go into the model and manually alter a single value downwards by £1.3 million – to make the QRA look better for the funders.

The analyst, seemingly untroubled by the ethics of the situation, adjusted the QRA in accordance with the instruction and reported back by email that he had "*pockled the spreadsheet*". (Brilliant verb, to pockle. I've found lots of uses for it since spotting it in the Inquiry report.) He confirmed it should help "*get the final result past CEC as I doubt they will notice what I have done.*"

They didn't notice.

As it turned out, the missing £1.3 million was a drop in the bucket against the magnitude of the overruns: three years late and £290 million over budget to deliver a line ending two kilometres short of the originally intended destination.

Add that one up.

No comment

Providing an approving body with commentary on an evaluation report isn't being helpful, it's courting disaster. The Australian Civil Aviation Authority found that out back in the 90s when they tried to procure an air traffic control system and failed dismally.

The initial evaluation report recommended Hughes Aircraft Systems International ahead of the second-ranked bidder, Thomson Radar Australia Corporation.

The report to the board was accompanied by a note with commentary from the CEO, which supported the recommendation but asked the board to note management's concerns about the risks of the Hughes proposal, in particular in relation to software development. The note didn't point out that the evaluation team had taken those risks into account but still recommended Hughes.

That was all it took. The board was completely spooked. The hare was off and running: the Hughes proposal was risky and the Thomson proposal was not. The board rejected the recommendation and sent management off to have another go.

Management appears to have taken this as an instruction to find evidence that the Hughes proposal was risky. They collected information on the relative numbers of Source Lines of Code, or

"SLOC", in each software development proposal. Thomson had a lot more SLOC than Hughes. The team seized on this as evidence the Hughes proposal was more risky.

A software expert, brought in for confirmation, instead advised the differences in the numbers could be expected because Hughes would be developing software from scratch and Thomson would be reusing existing software: customisation of existing software uses more SLOC. The difference in SLOC didn't necessarily say anything at all about the relative risk.

Oh dear. The expert's report had given the "wrong" answer and was ignored.

The board was then presented, appropriately enough on a Friday the 13th, with:

- the original summary of the evaluation, which still showed Hughes in front on an analysis that included an assessment of software development risk;
- a comparison table of the SLOC differential, minus the expert report that would have shown it was misleading; and
- a paper recommending Thomson on the basis the Hughes proposal was high risk.

Unsurprisingly the board, none of whom knew much about software, decided in favour of Thomson. The decision was set aside after an external review and the procurement process had to be re-run.

Not all commentary results in such a travesty. But the evaluation report should be a complete document providing a balanced view of all the relevant factors and signed off by the evaluators.

Adding commentary introduces bias and effectively amends the report unilaterally.

Any issues should be raised at the draft report stage, so the evaluators can consider them and decide whether the draft needs to be amended.

Never waste a near miss

The investigation of the collapse of Baltimore's Francis Scott Key Bridge in March 2024 is still a work in progress. One thing is fairly obvious. Any replacement bridge will have greater protection for the piers against being run into by a large ship.

And yet, at the time it was opened in 1977, the design may well have been adequate. Ships were smaller and less frequent, and the chance of one causing catastrophic damage to the bridge was accordingly very much lower. When a ship hit a bridge fender in 1980 the fender had to be replaced but the bridge itself was fine.

Boiling frogs spring to mind. The risk has obviously been increasing, and at some point the risk became high enough to justify the expense of adding more protection to the piers. When? That wasn't considered in the Marine Investigation Preliminary Report, but with the benefit of hindsight, we can surmise it was well before March 2024.

What Baltimore really needed was a near miss – not a quiet, sweep-it-under-the-carpet near miss but a highly visible, heart-stopping, OMG-that-was-close moment – that would have unlocked the will and the funding to upgrade the bridge protection.

It didn't get it. But perhaps it can provide one for the rest of us. How many bridge owners around the world stared at the video of the bridge collapse and thought "Shit, could that happen to us?"

Yeah, it probably could. Maybe this is a good time to seek funding for an upgrade.

Wrangling the 80/20 principle

The 80/20 principle is an excellent guide to prioritisation. But the Minister probably doesn't understand the maths.

Broadly, actions and outcomes are rarely in a 1:1 relationship. 20% of your actions will deliver 80% of your results. 20% of your customers will produce 80% of your profits. The ratio may not be 80/20 – it could be 70/30 or 99/1, but it is still a good idea to target your efforts onto the things that will make more of a difference.

You do have to be a bit careful when you go looking for funding.

If you say, "This project will eliminate the cause of 50% of delays on the rail network", it is more or less guaranteed that what the Minister will hear is, "This project will eliminate 50% of delays on the rail network."

Not the same thing at all.

If, for instance, 30% of train delays are being caused by starter motor failures, it might make sense to institute a starter motor replacement programme. That eliminates the cause of 30% of the delays. It will almost certainly not have reduced overall delays by 30%.

The trains that would previously have been delayed by duff starter motors have now been released onto the network, where they will for the first time have the opportunity to be delayed by defective brakes, points failures, suicidal cows and all the other exciting things that happen to trains on a complex network.

If you start with a network that is riddled with delays, it may not be until you get two or three or even more layers down in eliminating causes that you can really shift the dial on eliminating the delay itself.

By which time the Minister who mistakenly thought 30% of delays would vanish after the starter motor project will be baying for your blood.

Doesn't mean it was the wrong project. The 80/20 principle is great.

Just be careful how you communicate the expected results.

The clarity of hindsight

Why is it so difficult to see flaws in your own logic?

Back in 2012 the UK Department of Transport attempted to re-let the train operating franchise for InterCity West Coast. And failed. Comprehensively.

There were a number of things wrong, but the most critical was that bidders were required to provide a subordinated loan facility to ensure the successful bidder had adequate access to capital. The amount of the loan facility was calculated separately for each bidder, and was obviously critical to the bids. Would it be £60 million? £250 million? Zero? – it makes a difference.

The Department decided to calculate the amount of the loan facility for each bidder using a financial model which was not designed for that purpose. They recognised that if they disclosed the model to the bidders, it would be subject to challenge.

Their apparently logical solution? Since the model would be challenged if they disclosed it, they wouldn't disclose it.

Instead, the Department decided to issue guidance to the bidders, effectively saying, we won't give you the model but we'll tell you roughly how it works and give you a ready reckoner to give you an idea of the probable requirement.

Unfortunately, the guidance didn't match the model, including an elasticity factor that was out by more than 20% and a mismatch between real and nominal figures which meant the guidance underestimated the amount of the facility by about 50%.

To compound the problem, when the figures finally did pop out of the model, the Department decided they didn't like them anyway and proceeded to change them, which was not in accordance with the guidance issued.

And to cap it all, they exercised their discretion – the discretion which according to the guidance they didn't have – differently as between the two bidders. One loan facility was increased from nil to £40 million, and the other was reduced by £62 million, effectively imposing a penalty of £100 million against one of the bidders, Virgin.

The flaw in the original logic was then revealed. Virgin, which was the incumbent train operator, didn't actually have to have the financial model disclosed to them to realise that something had gone really badly wrong. Richard Branson complained, and when his complaints were ignored, sued. The competition was cancelled and the Department had to negotiate a two-year extension of the Virgin franchise just to keep the trains running while they sorted out the mess.

A less-flawed logic? If the original model isn't robust enough to withstand challenge from the bidders, use one that is.

Cheap and cheerful

The first railway suspension bridge, completed in 1855, was built over the Niagara River (part of the border between the US and Canada) below the Falls.

A problem for all suspension bridges is how to get the first line across the gap. Charles Ellet Jr, the original chief engineer (later replaced by John Roebling), decided to solve this by putting up a lightweight temporary suspension bridge.

Which of course created a chicken and egg problem: how could he get a line across the gap to support the construction of the temporary suspension bridge that would support the construction of the railway suspension bridge?

Proposals included firing cannonballs with the line attached and tying the line to a rocket that would then be launched across the gorge. Even in an age less conscious of health and safety considerations, these suggestions were somewhat alarming.

The final solution? Ellet organised a kite-flying contest.

The prize offered was $5 (a bit over $200 in today's money) to any boy who flew a kite across the gorge and secured the kite string to the other side.

Talk about community engagement. Youths from nearby towns apparently flocked in to participate in the contest, held in January 1848. The prize was won by 16-year-old Homan Walsh, the only boy who launched his kite from the Canadian side.

A feeder line was attached to the kite string and pulled across the gap, with the necessary heavier line then attached and pulled across by the feeder.

Solutions don't come much cheaper or more cheerful. But it worked.

Going the wrong way

Clients taking dumb risks on a megaproject tend to act like a macho cyclist going the wrong way down a one-way street. They expect everyone else to get out of their way.

The Queensland payroll debacle (see *Payroll problems* above) was a case in point. They knew from the start just how complicated the employment arrangements were (12 industrial awards, six different industrial agreements, 200+ allowances etc).

It should have been plain to the least IT-savvy observer that this state of affairs is a terrible starting point from which to implement a centralised payroll system. The cyclist was confronting a big red sign saying, "GO BACK YOU ARE GOING THE WRONG WAY".

Now fair enough, it's a persistent issue in government that some problems remain unsolved because they're just too difficult. No one can summon up the will or the capability to deal with the management and industrial relations challenges of fundamental change.

But if you aren't able to deal with the underlying issue, it isn't a good idea just to proceed as if it didn't exist.

With the local payroll officers being the only glue that held the existing fragile payroll system together, Queensland Health could have accepted that going to total centralisation with the new system in one big step was probably never going to work. If they had instead tried a staged process, retaining some local involvement, the project would have had a much better chance of success.

They didn't.

A pity. It would have been much less risky than hurtling down the middle of the road into the oncoming traffic.

A megaproject is not a side hustle

Social media is full of people saying everyone should have a side hustle. I disagree.

Do you want a side hustle to earn a bit of extra cash, or do you want to spend that time building your expertise so you can take a leading role on a megaproject?

Build mastery in your field and you get to be a part of awesome project teams delivering the infrastructure that powers and connects communities.

As a child growing up in Sydney, I used to look at the Sydney Harbour Bridge and think how amazing Bradfield must have felt, being able to point to it and say, "I did that."

As an adult, I learned he would have said, "We did that" – even the Bradfields and Brunels of this world don't do it alone.

But I also learned that being part of a team is part of the buzz. It is really energising to work with creative experts who can solve any problem you throw at them and to see all the disparate pieces come together to create something bigger than the sum of its parts.

A side hustle? Well, I get that extra cash is useful. But it's never going to deliver the Sydney Harbour Bridge.

Gaming the contract

There was a NSW construction contract that was essentially fixed price. However, the job required a lot of trenching for cabling, and the ground conditions were unknown.

Since it makes a big difference to cost whether trenches are dug through soil or through solid rock, the trenching element was separated out. In addition to the fixed price for the bulk of the work, the tenderers bid a schedule of rates for the trenching, so much per metre depending on the ground condition.

Pretty standard stuff. Except that for various reasons it became more convenient to do a large stretch of the trenching work in-house, and this was permitted under the contract. So a notice was sent to the contractor removing that element of construction from their scope of works.

Cue the sound of wailing and gnashing of teeth.

It turned out the contractor had deliberately low-balled the fixed price element of the works, expecting to make their profits from the somewhat elevated rates they had bid for the trenching work. Removal of the trenching meant they were doing the rest of the project for slightly less than nothing.

Ouch.

Quotas are a double-edged sword.

I came across one infrastructure organisation recently which has made it a requirement that all procurement evaluation panels be 50% female.

Good concept. Diversity generally gives better outcomes.

But the organisation is heavily male-dominated, so only 20% of the candidate pool is female. The effect is that the women are having to sit on more than twice as many of these panels as their male counterparts.

Do they get brownie points for that at bonus time? Or is it just an administrative burden that makes it harder for women to get ahead in an already male-dominated organisation?

I don't know the answer. But I suspect the procurement policy was adopted without anyone asking the question.

PART 4

First mover advantage

When it comes to recruitment for roles on a megaproject, the public sector client has one huge edge over the private sector contractor.

No, I don't mean their recruits have a higher chance of making the Honours List, even if it's true.

And it certainly isn't the ability to pay more. The public sector does better now than it has done in the past, but it is still pretty much true that at the professional level, the private sector pays more.

What the public sector does offer is certainty. The private sector has to bid for the contract. A bidder has no surety even of making the shortlist. If they do make the shortlist, and invest the necessary months and millions to put together a first class bid, there is still no certainty the bid will be successful. It can be a long and dispiriting road when you walk away with nothing.

The client doesn't know who the successful bidder will be either, but they do know there will be one. Whichever bidder gets the guernsey, the client will still need the same team, and the better the team, the better the likely outcome.

The client will also know the scope and timing of the contract before the bidders do. It's a splendid opportunity to get out there and start recruiting before anyone else is in the market for talent. There are plenty of great people out there who prefer the certainty of government work to advising private sector bidders who may not pay them at all if the bid is unsuccessful.

Too often, the public sector squanders the opportunity. You only get first mover advantage if you move first.

Frank and fearless. Not.

The Jasper Inquiry by the New South Wales Independent Commission Against Corruption back in 2013 examined tenders for valuable coal exploration licences. It found the relevant Minister had requested an extension of the period for submission of expressions of interest to allow additional bids, even though the period had already expired.

There was a probity auditor in place, an *"experienced public servant"*. He drafted a recommendation against the extension but withdrew the objection on being advised the Minister had already made up his mind on the issue.

The whole thing later blew up in everyone's face and the Jasper Inquiry found the Minister had made the decision corruptly to favour a third party. The report was challenged and it all got very messy.

It isn't always easy to raise an objection to a Ministerial decision. Doesn't mean you shouldn't do it.

Distorting the investment decision

Government subsidies are a commonly-used form of incentive. What they incentivise people to do is to spend other people's money.

Hopefully on something useful.

When the federal government offers subsidies to state governments for infrastructure, it enables projects to be built that otherwise could not be afforded. The Roads to Recovery program, for instance, is currently funding road projects in local council areas across Australia.

When it comes to megaprojects, the scale of the subsidies can have a significant impact on the investment decision. Adelaide built a desalination plant in 2010, partly funded by a federal government grant. It now delivers about 5,000 megalitres a year, well below its maximum capacity of 100 gigalitres. The Productivity Commission complained the decision to make such a large investment in supply relative to demand had been distorted by the federal grant of $328 million.

What did they expect? The whole point of a subsidy is to distort the investment decision.

And you never quite know if it will be distorted in the way you intend.

Did the state government ask, "Should we spend $328 million of federal money to build a bigger desal plant or is there a better use for the money?". Or did they ask, "Should we spend $328 million of federal money to build a bigger desal plant or see that money redirected to NSW or Victoria?"

Um.

As Paul Keating famously said, "Never stand between a premier and a bucket of money."

A CEO in conflict

The thing about a conflict of interest is not just that one of the conflicting interests will lose out to the other one. It's that, without specific intervention, the person with the conflict gets to pick which one. (Professional ethics would often suggest withdrawal from both, but not all people with conflicts of interest are professional or ethical.)

And it's all too easy for these things to slip by unnoticed – until the moment when the project goes totally pear-shaped, runs three years late and wildly over budget, and the Commission of Inquiry starts asking awkward questions.

If all had gone well, probably no one would have noticed the chief executive of the project manager for the Edinburgh Tram project was a member of the Remuneration Committee that determined his own bonus. Possibly he considered the issue was adequately dealt with by his decision not to attend the meeting where the bonus was determined. Alas, as the Commission of Inquiry report pointed out, this was sadly undermined by the fact that he was the one to propose what level of bonus opportunities should be available to him.

The chair of the organisation might have been expected to cavil at the arrangement, but (contrary to Cadbury and Greenbury guidance on corporate governance) the chair and the chief executive were one and the same person. Surprise! The chair did not object to the chief executive recommending his own bonus.

If all had gone well, no one would have noticed. True. But without good governance, things were never going to go well.

How long is a piece of string?

One of the more frustrating tasks of a megaproject director is to find good answers for people who don't understand their own questions.

I have a vivid memory of a luckless rail engineer trying to answer a parliamentary select committee member who had just asked how much time contingency was in the schedule for a new rail line.

How to explain just how meaningless this question was? For a small project, you do keep time contingency in the schedule. It should take six weeks, but something will go wrong, so let's allow eight. Two weeks contingency. Straightforward enough.

But for a megaproject? No way. The interdependencies are far too complex to allow anything so simple. Yes, there will be a bit of time contingency in the schedule. But the overwhelming bulk of your time contingency is sitting in the budget.

If a task that takes a week falls behind by a day, you don't pull a day's time contingency out of the schedule. If the task is on the critical path, you pull some money out of the budget and pay the gang to work an extra shift to catch up.

So how much time contingency was in the budget? The engineer struggled to communicate the length characteristics of this piece of string.

Some activities, such as laying long stretches of track, are relatively easy to accelerate by putting resources into working multiple sites. Others, such as pouring concrete footings for a bridge, take as long as they take and no injection of money will make much difference. The amount of time by which you can accelerate the project can depend as much on the creativity of the schedulers as on the funds available. And it changes every day as activities that could have been accelerated are completed.

The engineer totally failed to get this across to a sceptical select committee, whose members (with not a single STEM qualification between them) left the hearing convinced the project had no hope of recovering the six months delay caused by bad weather in the initial stages.

I still don't know how much time contingency was in that project's budget. But the rail line was delivered on time and with no cost overrun, so it must have been enough.

The other megaproject

The scale of megaproject construction can obscure the fact that the organisational change required to utilise the new infrastructure may well be a megaproject in itself. Ireland seems to be finding this out the hard way.

A huge new children's hospital is being constructed in Dublin, which will replace the three existing children's hospitals around the city. It is being described as the first "public digital hospital", so its information management is intended to be a big improvement on the current set-up.

In 2015, the new hospital had an estimated cost of €650 million and a completion date of 2020. For which now read €2.24 billion and 2025. Large as those figures are, they aren't currently the top concern.

In September 2024 the Irish media got hold of a confidential KPMG report prepared the previous April examining readiness for the transition. Or perhaps more accurately, lack of readiness for the transition.

Round about the end of 2025, the three old hospitals will have to migrate to the new one. That's not just a physical migration, which would be sufficiently challenging, it's a migration to new digitalised working methods. Plus it's a merger of three separate organisations with "marked differences in culture and organisational maturity". That's pretty mega.

But it hadn't been given the resourcing and focus you would expect on a megaproject. According to RTE, the Irish public broadcaster, the KPMG report reveals "gaps in senior leadership teams, a sub-optimal clinical governance structure, [and] a lack of clarity around the future operating model and resourcing of the new hospital."

None of which augurs well for a successful transition to operations.

There may still be time to put in a full, dedicated project management team and prevent the transition being a complete shambles. But the clock is ticking.

The value of an independent check

In 1876, the engineering firm of Sir Francis Fox was asked to supply plant for a company in Chile with a silver mine. It was quite an undertaking.

Sir Francis told the story in his book, *Sixty-Three Years of Engineering*, published in 1924: "The mine was high up in the Andes, so that every part had to be taken up by mules. Consequently the engines, boilers and every subsidiary part had to be so designed that no piece should exceed 3cwt. The shaft was 10 degrees from the vertical, and the cage was to run on inclined rails. There were winding engines for the shaft, hauling engines for the underground planes, pumping engines, boilers, guides, pulleys, ropes, wagons, and also the necessary buildings – all had to be perfectly complete to the last screw. We had it all put together at Messrs. Appleby's works at Leicester; tried in steam and tested, then carefully packed, shipped, paid for and our work was done."

Nothing further was heard from the customer. About three years later, Fox by chance heard it had been wholly unable to use any of the equipment.

The original order and specification had arrived in Spanish. Fox had declined to attempt a translation and asked the customer to provide one. Alas, the translator had made an error and accidentally doubled the dimensions. Nobody checked his work.

The entire plant, so meticulously designed and tested, was useless.

I wonder if it was before or after carting the pieces up to the mine that the customer realised.

Kicking the can

You're in a major contract dispute. The negotiation period is about to expire, forcing you into third party adjudication. Should you stop the clock?

You can't do it unilaterally, of course, but the parties will often agree to extend the negotiation period, meaning neither party can trigger third party adjudication until a later date.

That isn't always a good idea.

The time limit in dispute resolution clauses is there for a purpose. The longer an issue remains unresolved, the harder it is to find a good outcome. Positions become entrenched, facts become harder to establish, costs escalate, and the project schedule may crumble under the weight of the uncertainty.

Okay, megaprojects are complex, and they can give rise to complex disputes. You not only have to reach agreement, you have to document it carefully to avoid misunderstandings and unintended consequences. To do that well, you may need more time than is allowed in the contract.

In that case, extending the specified negotiation period is fair enough. No point in going to court if the substantive dispute has already been resolved.

But what if you have made no progress in resolving the dispute? Do you really have any reason to think the situation will be any better a month from now?

If you're just kicking the can down the road, don't.

Do you really need to know?

Bid documents for a megaproject are usually enormous. Bidders literally spend millions putting them together.

Does all of this stuff add value?

Possibly not.

When you ask bidders to provide information, there is one simple rule: if you do not know what you are going to do with it, do not ask for it.

On one public sector project, the team came up with about 100 returnable schedules of things they wanted the bidders to tell them.

This seemed excessive, so they were then asked to provide, for each schedule, a single side of A4 paper explaining what the schedule was, the benefits of having the information and how the schedule would be incorporated into the contract.

After this single intervention, the number of returnable schedules was down 30 per cent within a week.

Always tempting to ask for more information. Not necessarily useful to have it.

Alternative uses for airports

Airport projects give rise to surprising opportunities for ingenuity.

Unfortunately, this is because two out of every three airports operate at a net loss. My attention was drawn to this by Paul Hooper's chapter in the book *White Elephant Stampede*, which provides case studies of some of the more egregious white elephant projects perpetrated by governments over the years.

I was intrigued by the alternative uses found for redundant airports. (Did anyone else suffer the high school science lesson where the teacher tries to inspire creativity by making you come up with 10 alternative uses for a brick? Number one on everyone's list was using it to bash teachers who set stupid assignments. I digress. Airports.)

Montreal's Mirabel airport opened in 1975. It was a flop in its intended role as an international passenger airport, although it now does quite well with cargo. Alternative uses attempted in the meantime included:

- Racetrack (nice try, didn't last long)
- Film location (bit better, credits include The Jackal, Sidney Poitier's last film)
- Manufacturing base for Airbus Canada (successful, still in use).

Sri Lanka's Mattala Rajapaksa International Airport (built with Chinese money near the President's home town, say no more) came up with more exotic alternative uses:

- Terminal building – rice storage
- Runway – Tourist attraction, see the wild elephants (not intentional, 300 soldiers and police officers employed to drive off the elephants)
- Access road – Drying area for pepper harvests

Alas, St Helena (the island in the south Atlantic where Napoleon was imprisoned) proved entirely lacking in such ingenuity. With not even a B-movie credit to its name, the airport was taken over by the government in 2016 and dubbed by the British press "the world's most useless airport."

Western Sydney International Airport is due to open in 2026. Let's hope it doesn't need to come up with any alternative uses.

It's an ill wind ...

I remain grateful to Britain's Nuclear Decommissioning Authority for making a complete horlicks of the 2014 procurement of the £6.2 billion Magnox decommissioning contract.

I was at the time engaged in writing my first book, *Procuring Successful Mega-Projects: How to Establish Major Government Contracts Without Ending up in Court*. The NDA very kindly got itself sued by the losing bidder in the same week I submitted the book proposal to the publisher, thereby demonstrating the extreme topicality of the subject matter. My proposal was accepted immediately.

The UK government was probably less grateful.

The later Inquiry found that: "*In many respects, Magnox was a well-run procurement and appeared to have the critical components for successful delivery. These included a tried and trusted procurement model (competitive dialogue) that was understood by the market; a multi-level governance structure with appropriate stakeholder representation; market engagement; appropriate policies, risk identification and regular reporting; a seemingly well-resourced team; the engagement of external advisers and independent internal and external assurance.*"

All those good things gave rise to a misplaced confidence in the evaluation process. The NDA had permitted the adoption of a thoroughly unwieldy evaluation methodology. I say "permitted the adoption of" rather than "adopted", since the thing seems to have grown organically rather than been intentionally designed. It had more than 2,800 requirements to be individually scored, including more than 300 mandatory criteria.

The court found various problems with the evaluation, including that the winning bidder should have been eliminated because it failed to meet two mandatory criteria: the evaluation had been fudged by the NDA.

Getting only two wrong out of more than 300 may sound like a fantastic result. Unfortunately, a bidder can't be a little bit eliminated.

The NDA ended up paying losing bidders almost £100 million in compensation.

Win-win? Or lose-lose?

It seems some contractors can be more creative at ripping off subcontractors than they are at delivering the contract.

I came across an interesting example recently. The contractor has a not uncommon arrangement whereby they receive payment from the client if there is an increase in material quantities.

What is less common (at least I hope so!) is that they also have subcontracts which allow them to claim payments from their subcontractors if … there is an increase in material quantities.

Yep, that's right. If quantities increase, they can claim from the client AND from the subbies.

Not exactly incentivising the contractor to achieve savings in material quantities.

Somehow, I doubt that's what the client had in mind.

How short is a shortlist?

There's lot of leeway, but don't believe anyone who tells you that "probity" requires you to have at least three fully-priced bids.

It is true in general that having a number of bids makes it harder for a corrupt official to award a contract to a mate and makes it more likely you will get genuine price competition. It is not true that a process involving fewer than three bidders is necessarily invalid or corrupt.

For a megaproject, bidders need to invest millions in the tender process to create a fully-priced bid. They will not make the investment unless there is a reasonable chance of winning the contract. When you are shortlisting bidders to be invited to give you a price, three is more likely to be a maximum than a minimum, and you may well generate fiercer competition with just two bids.

Sadly, the public sector tends to put off making difficult decisions and eliminating a competent bidder is always going to be a difficult decision. But let's face it, eliminating all but one of the bidders is the whole purpose of the procurement process. A long shortlist just postpones the problem.

It also puts up the cost to industry (and the client, but at least they deserve it for saddling themselves with a long shortlist to evaluate) and increases the risk of challenge, since you end up with a longer process and more bidders who have sunk more cost into the bids.

All in the name, really. It's called a shortlist because it's supposed to be a short list.

Training the monkey

When you're planning infrastructure projects, where should you start?

Annie Duke, author of the excellent book, *Quit: The Power of Knowing When to Walk Away*, reports that X – that's Google's moonshot company, not the rebranded Twitter – has a way of figuring out what to spend money on developing.

It's a bit weird, as you would expect from Silicon Valley. The idea is that if you could train a monkey to juggle flaming torches, and installed it on top of a pedestal in a public park, people would pay a shedload of money to come and see it. There are just two things you need to do. Train the monkey and build the pedestal.

The almost overwhelming temptation is to get stuck in building the pedestal, because hey, we know how to do it, we can estimate the costs fairly precisely and we can get started before the election. And yes, you will end up with a very nice pedestal. But no one is going to pay money to come and see it.

So the X approach is that you should ignore the pedestal and start out by training the monkey. If you're successful, adding a pedestal won't be a problem. If you're not, you won't need a pedestal.

From a megaproject point of view, when you are structuring your infrastructure procurement, the place to start planning is not with whatever you did last time that you felt comfortable with, even if it worked. Just because you know how to do it doesn't mean it's the best thing to do this time round.

The best place to start is with the biggest risk, because on a megaproject the biggest risk is often humongous. How can you structure things to cut that risk down as far as possible? Not just shove the risk off onto the contractor, but actually reduce the risk? That's what will add the most value.

The nature of the risk will vary. Maybe it's getting planning consents. Maybe it's a specific engineering challenge. Maybe it's the difficulty of getting a skilled workforce. If it's a light rail project it's probably the utilities relocation. Whatever. Getting that bit right transforms the whole project.

It's unlikely your situation will be as extreme as California's high speed rail project, which went full steam ahead without first solving the engineering challenge posed by the Tehachapi Mountains at the Los Angeles end and the Pacheco Pass at the San Francisco end. Thirty-five billion dollars later the project is still in progress, building a railway from somewhere nobody's heard of to somewhere else nobody's heard of, but the odds are overwhelming it will never get to either Los Angeles or San Francisco at all.

But you can still add a lot of value by tackling the biggest risks first.

Or, as X would put it, start by training the monkey.

The Colombiera Bridge

Legendary stupidity is sometimes just that – a legend.

One such is the story of Intermarine, the Italian shipbuilding company that built four large ships, as told by Stephen Pile in *The Return of Heroic Failures*: "Only when the huge craft were completed did the builders recall that their shipyards were connected to the sea by the River Magra upon which nestled the attractively minute Colombiera Bridge. Not one of their new vessels was able to pass underneath it. Intermarine offered to knock down the bridge and rebuild it, but the local council refused and the people of Ameglia gathered round to admire their new navy."

I found it hard to believe a shipbuilder would have been quite that stupid (there is, of course, a lot of stupidity about, but even so ...). So I did a bit of digging. Google found AskMetaFilter for me, where the question had been posed and answered.

Sure enough, it seems the builders knew perfectly well the bridge was a problem. When the shipyards were first erected in 1970 the local mayor had promised to allow modification of the bridge later. In 1976, when tendering to build four fiberglass minehunters for the Italian navy, Intermarine applied to ANAS, the Italian road authority, to modify the bridge at the same time. The request was approved, so all should have been well.

Unfortunately, no local consultation was carried out, and the outcry prevented modification of the bridge. Intermarine then looked at all sorts of options for getting the ships past the bridge, including the possibility of digging a new canal, but nothing looked practicable.

Finally, with finished and near-finished ships idle in the shipyard, the Ameglia council voted to approve the modifications in January 1983, and the new Italian government at last provided approval at the end of the year, with a raft of conditions to minimise impact to the local community. The modifications were completed in May 1984.

The date seems to have been driven less by the need to deliver the ships and more by being just in time for the Colombiera

Bridge to take its place on the route of the 1984 Giro d'Italia, but perhaps that's just me being cynical.

Viva l'Italia.

A public sector scheduling problem

How do you get authority to award the contract?

In a private sector company, authority comes from the board of directors. A resolution of the board is normally all it takes to authorise even the biggest contract. There may be a need for shareholder approvals in very special circumstances, but for all practical purposes the board is god.

The public sector offers a greater variety of deities. Government agencies may have unique powers and structures since they can be created by specific legislation rather than established under the regular laws governing companies.

Executives brought in from the private sector are often bemused when they run into this kind of thing. They sit in meetings muttering – occasionally even shouting – "JFDI, JFDI". (That's private sector speak for "just do it".) The arcane intricacies of identifying sources of authority in a public sector environment may seem bizarre.

Sometimes it is, in fact, bizarre, but it's important because you have to build it into the project schedule.

The $3 billion contract for the M7 Clem Jones Tunnel was let in 2006 by Brisbane City Council. Brisbane City Council was the largest local government authority in Australia, with 26 wards, 27 councillors and – this was the kicker – no legal capacity to delegate the award decision.

Obtaining authority to award the contract was never going to be achieved by just turning up at a council meeting. Even for Australia's largest, a $3 billion contract was not a routine decision, and the debate would not be enhanced by hecklers from the public gallery.

The final deliberations took about three days, and to preserve the confidentiality appropriate to such a decision, the councillors agreed to be sequestered for the duration in a hotel in Brisbane's CBD.

Not the kind of thing you want to be setting up without notice. Schedulers take note.

The romance of project finance

I recently re-read with pleasure the Nevil Shute novel, *Ruined City*, where the inevitable romance emerges in a fascinating clash between personal and professional integrity.

Henry Warren, the hero of the book, is an investment banker in the City of London during the Depression. He has a well-earned reputation for honesty and integrity in his business dealings.

Chance puts him in hospital in fictional Sharples, a once-proud northern English town which has had the heart ripped out of it by the failure of the shipyard that was once the biggest local employer. He hears the town's story from the hospital almoner, a young local woman. To save the town, he acquires the old shipyard, then concocts a dodgy deal in the Balkans, using a specially commissioned green silk umbrella with a jewelled handle to bribe a local official, who places an order for ships with the Sharples yard.

To finance it, Warren sets up a company and issues preference shares, taking 25% himself and getting the rest of it away to investors with a profit forecast he knows to be wildly exaggerated and security from a Balkan government which is in no way secure. Only the strength of his personal reputation carries the day: "If it was anybody but Warren," says one of the investors, "I'd have nothing to do with it."

The shipyard is resurrected and Sharples is re-born. The investors are not so lucky, losing all their money when the Balkan government falls. Warren, without hesitation but also without remorse, pleads guilty to issuing a false prospectus and goes to gaol for three years. At the end of his sentence, he returns to a now-prosperous Sharples to the acclaim of the workforce. And, of course, marries the almoner.

Shame about the investors.

Risk mitigation can be fatal

Actions undertaken to mitigate one risk can create a bigger problem than the risk they were trying to manage. I call it "iatrogenic risk".

"Iatrogenic" is an adjective used in medicine for diseases or injuries caused by medical examination or treatment. As far as I know, there is no word for the equivalent in ordinary risk management, so I borrow the medical term.

There is a sad illustration of iatrogenic risk from the first world war.

The British navy had introduced cork lifejackets to reduce deaths from drowning. The lifejackets, which had a stiff cork collar intended to support the head above water, worked splendidly for a sailor washed off the deck of an ordinary yacht.

It occurred to no-one until the day it happened that abandoning a frigate sinking under enemy bombardment might involve leaping into the sea from a deck that could be 100 feet above the water. The collar designed to support the neck instead broke it on impact, killing the sailors instantly.

The hidden costs of getting it wrong

A good reason for getting the procurement of big contracts right is the cost and the suffering caused by the proliferation of public inquiries if you get it wrong.

The InterCity West Coast £5.5 billion rail refranchising back in 2013 was a prime example.

As soon as the preferred bidder announcement was made, Virgin, the incumbent franchisee and a losing bidder, brought an action for judicial review. The Department for Transport took one look at the evidence and promptly cancelled the competition.

Sam Laidlaw, the lead non-executive board member conducted an inquiry for the Department for Transport. He looked at what had happened, with particular reference to technical flaws in the process; roles and responsibilities; and arrangements for review and quality assurance.

There was also an HR inquiry, which was not published. Three people were suspended while the inquiry was carried out, one of whom challenged the suspension in the High Court. More fun and games.

The National Audit Office did its own review. It estimated that the Department for Transport had spent or would spend £1.9 million working on the cancelled competition and £2.7 million on professional fees relating to the judicial review.

Richard Brown, the Chairman of Eurostar, was asked to report on the implications for the rest of the rail franchising programme.

The Transport Select Committee also joined the free for all and demanded even more investigations.

The Department ended up spending £4.3 million on the various public inquiries associated with the fiasco, in addition to the £1.9 million on the cancelled tender process and £2.7 million for the judicial review. Not to mention the suffering of managers being grilled in one inquiry after another.

Would have been cheaper and simpler to get the procurement process right in the first place, don't you think?

Finding where the bodies are buried

Providing a "dial before you dig" service is – or ought to be – one of the key functions of the General Counsel.

Let's face it, when big trouble strikes, the General Counsel is usually the first port of call. They know which departments have trouble with their contractors, which employees have been investigated for what, who writes aggressive emails in disregard of courtesy and common sense, who runs for cover at the first hint of a problem, and who picks up the pieces.

Project managers are often shy of calling lawyers, but the expertise of the in-house variety is a valuable resource when you're at risk from the concealed cadavers tucked away in the business. Call them before breaking new ground.

They may not be able to tell you all they know, because much of what they do is confidential, but their recommendations are usually worth listening to.

Ten business days

One small clause in the contract, lots of big (and mostly good) consequences.

When Sydney bought 78 Waratah trains under a PPP entered into in 2006, one of the most significant provisions in the contract was a restriction on delivery.

When a train achieved Practical Completion, it would then be ten business days before the next train was permitted to achieve Practical Completion. (The period for the first six trains was 15 business days in case there were teething troubles.) That gave a delivery period of about three years.

This simple provision had huge impacts.

- It simplified the planning for the introduction of the trains, reducing the risk of driver training and network projects getting out of synch with the train delivery.
- It forced contingency into the schedule. Train manufacturers work at different rates, much faster now than 20 years ago, but it was certainly practicable at the time to manufacture at the rate of one car per working day. The restriction created almost eight months of contingency, even without taking into account the possibility of weekend and holiday working.

- It prevented excessive risk taking by the tenderers. They would inevitably be tempted to bid the fastest possible rate of manufacture to reduce costs and secure the contract. If delivery ran into trouble, they would be left with huge financing costs and no ability to recover from delay, potentially sending them bust while the trains were still being manufactured, a nightmare scenario for the client.
- It simplified tender evaluation. Since the fastest delivery rate permitted was clearly practicable, all the tenderers adopted it, so there was no need for complicated comparative risk assessments associated with different delivery rates.
- It prevented the contractor delivering a number of trains at the same time, avoiding the client being faced with a choice between skimping on final checks and enabling the contactor to get away with unfixed minor defects, or delaying acceptance of the trains and making delay payments to the contractor.
- It did NOT secure the lowest possible price: faster delivery rates would have lowered contractor costs and therefore reduced the headline price.

But one of the most significant impacts was perhaps the least obvious. The contingency forced into the schedule belonged to the client.

If the contractor was late and wanted to make faster deliveries to catch up the schedule, it was up to the client to decide whether to allow that. If things were going smoothly, and the contractor wanted to improve profits by bringing forward delivery and associated payment dates, it was up to the client to decide whether to allow it. Throughout the whole of the delivery period, the client had a valuable bargaining chip to hand in its dealings with the contractor.

And when you're trying to secure value for money on a megaproject, that's no bad thing.

If you see something, say something

That doesn't just apply to bombs on public transport. If, for instance, you become aware people are about to spend hundreds of millions based on incorrect assumptions, you might want to point that out.

It is not, of course, always so easy. The Edinburgh Tram Inquiry, convened after a supposedly fixed price construction contract had a massive cost overrun, published its report in 2023. The Inquiry found an attempt had indeed been made to warn the City of Edinburgh Council (CEC), which was funding the project.

The contractor's representative, a Mr Walker, was concerned whether CEC fully understood the cost to them would increase after contract signature: the assumptions on which the price was based were already known to be wrong and it was the client rather than the contractor who would bear the "risk" (i.e. certainty) of the price being higher. He asked the Executive Chairman of the client's project management agency, **tie**, to confirm that CEC was aware of the price increase and was told they had been informed.

Mr Walker drafted a follow-up letter to confirm this. Possibly he foresaw that, as indeed happened, the Executive Chairman would later deny the conversation had ever taken place. (The Chair of the Inquiry "preferred the evidence of Mr Walker".)

The draft letter would have given anyone pause, since it pointed out that "*the woefully inadequate progress of the utility diversions were dramatically going to affect the price by a significant number of tens of millions*". Alas, the letter was never sent. Mr Walker's boss considered it would "spoil the working relationship".

Well, yes, something as forthright as that probably would have spoilt the relationship. But should that really have been the primary consideration?

PART 5

Independent. Not.

Independent reports are an essential tool of governance. Unfortunately, they don't always say what you had in mind.

And it's the response to that which shows just how effective your governance really is.

When concerns were expressed about the letting of two contracts as part of the London Garden Bridge project, TfL asked its internal audit team to conduct an audit. The purpose, according to the subsequent 2017 inquiry, was *"to provide assurance that the procurements had been made in accordance with procurement regulations and approved procedures, and were open, fair and transparent"*.

They hadn't been: *"Our audit identified a number of instances where the procurement deviated from TfL policy and procedure and OJEU guidance ... taken together these adversely impacted on the openness and objectivity of the procurements."*

And, *"The nature of the findings from this audit we believe increase the risk of legal challenge by the unsuccessful bidders for both contracts. It is the informal contact between TfL and individual bidders that has had an adverse impact on the transparency of each procurement."*

What was the response? Well, the Inquiry found those sentences in drafts of the report. They didn't appear in the final version.

The final version was edited to focus not on whether the procurement process had been properly conducted but on value for money. Not that that was totally helpful.

The best they could come up with was: *"The audit did not find any evidence that would suggest that the final recommendations did not provide value for money from the winning bidders."* Whoopi doo.

The revised version was distinctly less embarrassing than the original, but did not, alas, do much for project governance.

The London Garden Bridge was never built. It still managed to suck £43 million from the public purse.

Assessing the evidence

When you write a prospectus, you have to have evidence for what you say. The lawyers will create a separate document for the record which takes every statement in the prospectus and notes the source of the information or claim. While people will pay particular attention to things like profit forecasts, the checking process applies to absolutely everything.

My first railway job was to write the prospectus for the flotation of Railtrack, predecessor to Network Rail. I was assured that Railtrack had inherited more than 24,000 miles of track from

British Rail. That was the number in all the brochures from British Rail days, so it must have been true, mustn't it?

We decided aging PR brochures were insufficient evidence and probed a bit deeper, working it out line by line, region by region. Every time a new tranche of verification was completed, the number dropped. And dropped.

The line in the draft prospectus went from "more than 24,000 miles" to "about 24,000 miles". Sometime later it was "more than 22,000 miles". Then "about 20,000 miles." The number finally stabilised somewhere over 19,000 miles, which we decided was still near enough to sustain a prospectus statement of "about 20,000 miles", and that was what we ran with.

But it did make me wonder about the cost forecasts for track maintenance. When you don't know within 20% how much track you've got, how good can your maintenance forecasts be?

Mandatory criteria

Mandatory means mandatory. Except when it doesn't.

Setting assessment criteria before carrying out an assessment is a no-brainer, or ought to be. The tricky bit is when the assessment against the criteria comes up with an answer the politicians don't like.

When Scotland ordered two ferries for delivery in 2018, it included a mandatory evaluation criterion – the bidders had to provide a Builders Refund Guarantee.

This guarantee is a feature of the standard New Build Contract of The Baltic and International Maritime Council, which is used throughout the shipbuilding industry. The contract stipulates that the full risk for the design and build remains with the builder throughout the construction of vessels. The buyer makes progress payments, but the builder provides a guarantee from a suitably accredited bank that the buyer will get their money back if the vessel is not delivered.

During the tender period, Ferguson Marine, a company which had only just emerged from administration, warned CMAL, the procuring agency, that they were having difficulties finding a bank to provide the guarantee, but said they would give it a go. On the strength of this, they were appointed preferred bidder.

Note that they were the only Scottish bidder, the last surviving shipbuilder on the lower Clyde and would probably plunge straight back into administration if they didn't get the contract. Dare I suggest these factors influenced this somewhat dodgy decision?

Alas, the decision promptly got dodgier. Almost as soon as it was appointed, Ferguson Marine said, sorry, no can do. They hadn't been able to find any bank that would provide a guarantee. (In view of the subsequent project history, you have to admire the perspicacity of the banks.)

Instead of throwing Ferguson Marine out, CMAL decided the guarantee wasn't mandatory after all. Astonishingly, none of the other bidders bothered to sue them.

Still a bad call. The project lurched from one failure to another. The government in desperation nationalised the shipyard in 2019, which didn't noticeably improve the situation. The first ferry finally entered service in January 2025. The other is still a work in progress. The spend to date is over four times the original budget and will go higher yet.

The project entry in Wikipedia is headed simply "Ferry Fiasco (Scotland)".

Says it all, really.

Bet CMAL wishes the moneyback guarantee really had been mandatory.

Another slice of cheese?

We're all familiar with James Reason's Swiss cheese model of risk, put forward in his book, *Managing the Risks of Organisational Accidents* (See *Figure 2: Swiss cheese risk model*).

The risk controls you put in place are like defensive walls against loss. Multiple controls give you defence in depth. The walls aren't perfect, but have holes in them like Swiss cheese. If you don't pay attention, the holes will get bigger. People stop bothering to follow procedures, don't fix broken alarm bells and generally slacken off. Sooner or later, the holes line up and the bullet with your name on it goes straight through the lot.

Figure 2: Swiss cheese risk model

It is very tempting to deal with the holes-lining-up risk by adding another control to form yet another defensive layer.

You see it all too often in the public sector. How many times have you seen a document that requires fifteen signatures before something can be approved?

Supposedly each signature is a check against something or other, but that's not much use if nobody bothers to define what.

There is a world of difference between "I have conducted the necessary investigations to confirm the proposed new electric train does not require the construction of additional substations" and "Yes, ok, you want to buy a new train, not my department."

One person, asked for a specific sign-off, will probably do a decent job. Their slice of cheese will be as hole-free as practicable.

The same person, asked to provide another signature on a document that has already been signed by fourteen other people, may not even realise their signature is there to provide a slice of cheese.

The more defensive layers you have, the less people think the layer they are responsible for is important and the more likely they are to let holes grow unchecked.

The lessons of the Swiss cheese model are that you need fewer and smaller holes, and you need to be vigilant against the holes getting bigger.

More slices of cheese? Not so much.

In medias res

"*Somewhat surprisingly the project has commenced delivery without knowing where it will start or finish.*" – Kerry Schott report into the Inland Rail project, January 2023.

Surprising? Yes and no.

The Inland Rail project is, according to its website, a 1,600km freight rail line that will connect Melbourne and Brisbane via regional Victoria, New South Wales and Queensland. So we do know the endpoints are Melbourne and Brisbane. Of course, the greater Melbourne area covers just under 10,000 square kilometres, and greater Brisbane covers more than 1,500 square kilometres, so there is scope for narrowing things down a bit.

But you could take the view – and somebody clearly did take the view – that with 11,500 square kilometres to play with, finding a location for two freight terminals is probably not an impossible task. So the decision to go ahead was not entirely surprising. Why waste time waiting for the terminal locations to be finalised when you could be getting on with the middle section while somebody works it out?

Sometimes that sort of thing works quite well. New rail and tram lines are often constructed in stages. The Northwest Metro in Sydney, for instance, was completed and open for business before the City Metro it recently connected to was fully planned and funded.

Sometimes it doesn't. The Inland Rail works in the middle are still going full steam ahead. Will they be worth the cost if the ends never arrive? Perhaps it's best to hope we don't find that out the hard way.

Real men don't do safety

It's an attitude that has killed a lot of people. Mostly men.

The English, blessed or cursed with a mild and rainy climate, have long treated the weather as a bottomless cup of small talk rather than a potential threat.

In the early twentieth century, in the days before the rolled umbrella and the bowler hat became the unofficial civil service uniform, carrying an umbrella was regarded by the English as a sign of effeminacy.

Which is why, when a group of English engineers visited the Panama Canal works before the outbreak of WWI, they enjoyed the blazing sunshine and, when a tropical downpour came on, were astonished to see that in an instant some hundreds of labourers each put up a large umbrella.

They waxed sarcastic on the subject.

General Gorgas, in charge of the works, was a lot less amused than they were. Possibly because he wasn't English. The labourers were highly susceptible to chill, and if after working in the blazing sun they suddenly got wet, fatal pneumonia could set in.

In those days bosses were less concerned about workers dying on the job, but the death rate from this cause had been a serious impediment to the works.

The General insisted on each man being equipped with an umbrella. Problem solved.

Real men use their PPE.

The negative side of modularity

Modularity is a great principle for the design and delivery of megaprojects. If you do it right.

The reason for its greatness is that repetition of the module allows you to learn and improve as you go. Flyvbjerg and Gardner give a number of good examples in their book, *How Big Things Get Done*.

What they don't mention, no doubt because it really ought to be blindingly obvious, is that if you just do the repetition without the learning and improvement, you don't get the benefit. Indeed, you not only don't get the benefit, you compound the disbenefit.

A friend of mine learned this to her cost when she decided to take up quilting and fixed on a design based on hexagons. She collected material, cut out the necessary number of hexagons, and only then discovered her template was out of true, so the pieces didn't fit together. The quilt was abandoned. Fortunately, it was a relatively cheap mistake.

With infrastructure projects, failure to learn can be costly.

To take a sadly not entirely imaginary example, train manufacture is a modular project: the same design is manufactured multiple times. But if the first train off a production line is too wide to fit next to the platform, it won't really improve matters if all the subsequent trains are equally off-spec.

If you are doing something wrong, doing it wrong a lot of times doesn't deliver the modularity benefit.

Proving a fair process

You may be running a fair process, but can you evidence it?

The losing bidder on a megaproject is inevitably going to feel hard done by. They have just spent millions – how could they possibly have failed to win the biggest contract going?

There's always a risk a losing bidder will be so convinced the process was unfair they will sue. So it's not enough that your process is fair, you need to collect evidence along the way to prove it. (A good probity auditor can come in useful here.)

On one New South Wales tender, the chief executive of the losing bidder sought an audience with the Minister to complain the process had been unfair.

The project director, invited and forewarned by the Minister, reviewed his collection of evidence and brought along a recording of the de-briefing meeting held with the losing bid team.

At the psychological moment he played an extract, an exchange with the tenderer's bid manager which went roughly as follows:

Question: "Do you understand why we decided to award the contract to the other party?"

Answer: "Yes we do."

Question: "Are you satisfied the process was properly conducted?"

Answer: "Yes, we are."

Question: "Do you have any complaints about the process?"

Answer: "No, we don't."

End of meeting. History does not record what the chief executive later said to his bid manager, but the challenge died in the Minister's office.

Defining success

The trouble with success fees is that you have to be really careful how you define success.

Construction contractors are not the only ones who follow the money. The last thing you need is a financial adviser with an incentive to close a poor deal.

The scandal of the treatment of postmasters has somewhat obscured the optimism around the Royal Mail privatisation in 2013 – it was supposed to a) herald an era of greater efficiency and b) pop some money into government coffers.

I'm not sure about a), and the National Audit Office definitely wasn't thrilled about b). One of their criticisms of the sale process was that the independent corporate financial adviser was only incentivised to achieve the sale, not to optimise value for the taxpayer. Since the shares rose 38 per cent on the first day of trading and remained 72 per cent above the issue price five months later, it is unsurprising the National Audit Office found the Government could have achieved better value.

One cannot, of course, blame the share price entirely on the structure of the financial adviser's incentive.

But you have to think a better definition of success would have resulted in a better outcome.

How many megaprojects is too many?

A very high profile action was brought for judicial review of the decision by the Secretary of State for Transport to award a £5.5 billion contract for the InterCity West Coast (ICWC) rail franchise in August 2012.

The case, brought by Virgin, the incumbent franchisee and losing bidder, was settled very quickly, because after one look at the evidence the Department for Transport cancelled the award and did a deal to extend the existing contract for two years while they sorted out the mess.

The various inquiries that followed found quite a number of things wrong, but the governance failures were particularly staggering. The official governance structure, which was in any event largely ignored, did not require the decision to award the contract to be considered either by the Executive Committee (ExCo) or the full board. And it wasn't.

Admittedly the board was only an advisory board – real power was with ExCo. The request for tender was issued on 20 January 2012 and the award announced on 15 August that year. But of the 26 ExCo meetings held between 10 January 2012 and 14 August 2012 there was reference in the minutes to rail refranchising on only three occasions and no explicit reference to the ICWC franchise process at all.

We're talking about a five and a half billion pound contract which was pioneering a new structure of franchise agreement. ExCo? Not interested.

ExCo was, of course, not merely sitting around twiddling its collective thumbs while the ICWC franchising process went down the pan. This was 2012.

High Speed 2 was mega. Thameslink was mega. CrossRail was mega. The Aviation and Road Strategy was mega. The London Olympics was humongously mega.

But if you can't find the time to consider a £5.5 billion contract, you've probably taken on at least one megaproject too many.

The privatisation problem

The thing about outsourcing a business to the private sector is that once you've sold it, you no longer own it.

In the early 1990s, before I gave up being a lawyer, I was advising eastern European governments on the laws required to establish a market economy after the fall of the Berlin Wall and the break-up of the Soviet Union. All the businesses had been government-owned and most of them were now being sold.

These governments didn't always cope terribly well with the concept that once they had sold something to the private sector, they didn't own it any more. It was offensive to them that where the business managers had once leapt to do their bidding, their requests were suddenly ignored and their instructions were unenforceable.

At the time, I thought these governments had difficulty with the concept because they were communists. With the benefit of another 30 years of experience, I think it was because they were governments.

Ministers are empowered by law to give directions to their departments that must be obeyed. The etiquette of the public sector is such that a Minister is rarely called upon to formalise directions in writing. This happens only when there is a specific legal requirement for a formal document of authority, which is not common, or where the department officials are so appalled by what the Minister is stubbornly insisting on that they refuse to proceed without a direction in writing, which is exceedingly rare.

Governments therefore often think public sector employees fall in with their suggestions as a matter of divine right rather than legal obligation and assume private sector businesses will be similarly compliant. They won't.

If you're doing an outsourcing contract, that doesn't mean you should attempt to replicate public sector constraints in the contract. Let's face it, the whole point of an outsourcing is normally to shed the shackles of the public sector.

But it's probably as well to make sure your Minister understands the basis of the new relationship established by the contract.

Otherwise your next job could be re-nationalising the business you just privatised.

Perverse incentives are everywhere

One of my favourites is the finding of the Dead Sea scrolls.

After the initial scrolls were found at the first of the Qumran Caves, excited scholars offered to pay the local Bedouin for any further scrolls they discovered.

Unfortunately, they promised a flat rate per find. It didn't take the locals long to discover that if they tore a scroll in half, they would be paid double. Tear the pieces in half, double again.

By the time the scholars realised what was happening, so many scrolls had been torn up that it took years to put them back together again.

I now can't remember where I came across the story, so I can't vouch for its provenance, but it's an object lesson in how careful you have to be in designing incentives.

Reference class forecasting

If ever there was a project that demonstrated the benefits of reference class forecasting, the Edinburgh Tram project was it.

Not because they used reference class forecasting successfully, but because they failed to use it and got egg all over their faces.

The issue was, as so often with light rail projects, utilities relocation. The Inquiry report, issued in September 2023, noted that the on-street section of the Edinburgh Tram project, from Haymarket to Newhaven, was approximately eight kilometres. The schedule called for utilities relocation along that stretch to be completed in eighteen months.

Reference class schedule forecasting requires you to look at comparable projects to see how long they took to do roughly the same thing. Obviously, if you aren't Bent Flyvbjerg you may not have a big enough project database to create a full reference class. But there's always Professor Google.

Two obvious light rail project comparisons were Manchester, where there was a two-kilometre on-street section, and Birmingham, where there was an 800-metre on-street section. In each case the works took about two years to complete.

So, the Edinburgh team couldn't possibly have guessed that eighteen months might not be enough time to clear eight kilometres. Totally unpredictable, right?

In the event, it took more than four years to complete six of the eight kilometres, which based on the comparisons may have been quite good, really. It just bore no relation to the schedule advised to the funders, whose fingers were badly burned as a result.

The evaluation spreadsheet of doom

It may come as a surprise, but the aim of the technical evaluation is not to come up with the world's most complicated spreadsheet.

We've all seen it. Every criterion divided into sub-criteria. And sub-sub-criteria. Etc. Each one separately scored and weighted. Hundreds upon hundreds of entries in the evaluation spreadsheet.

Okay, I'm exaggerating a bit. And it's true there is value in demonstrating that consideration has been given to every aspect of the bid submission.

But the purpose of a technical evaluation is to decide which bidder has the best technical offering. The more sub-criteria and sub-sub-criteria you have, the harder it is to distinguish between the bids. Even the wow factor of a splendid innovation may add only 0.01 to the overall mark when scored as a sub-sub-sub-criterion.

For each project, there will probably only be a handful of key items that should differentiate the bidders. If the scoring system does not allow the differentiators to make a difference, you can spend a lot of time filling in the technical evaluation spreadsheet and still come up with virtually identical scores.

Seriously unhelpful.

Treating bidders badly

Bidders for megaproject contracts are tough characters. They're not going to curl up and cry if you mess them around or make their job unnecessarily difficult.

But that's no reason to treat them badly.

In the InterCity West Coast franchise case, the UK Department for Transport originally advised the market it planned to issue the invitation to tender in May 2011. On the very day the invitation was to have been issued, the Department said it wasn't ready and put off the tender process indefinitely, eventually going to market in January 2012.

That didn't show a lot of respect for the bidders, who had taken on premises and bid teams ready to start work the instant the tender request was available.

There was no direct link to the problems that subsequently caused the losing bidder to sue and the competition to be cancelled, but the failure to give notice of the change put up bid costs and did nothing for anyone's confidence in the process.

If you can't be perfect, at least be consistent. Clients with a reputation for being straightforward and easy to deal with attract a better class of bidder and more competitive pricing.

A unique PPP?

When RailCorp (predecessor to Sydney Trains) decided back in 2004 to buy trains through a PPP, we thought it was a world first. A few places had bought whole railways through a PPP, with the trains as part of the package. And railway companies had been buying trains without a PPP for more than a hundred years.

But so far as we knew no one had ever used a PPP to buy trains that would be financed and owned and maintained and made available by the contractor but would be operated by someone else on a network over which the contractor had no control. (I later discovered London Underground had already done something very similar, but we didn't know that at the time.)

We did know it was tricky. The contractor would be responsible for train condition and maintenance during the operating period, but it would only get its hands on each train about once a month. In between times the trains would be whizzing around the network under someone else's control. Driven by people the contractor didn't train or manage. Running over track maintained to somebody else's standards. Rained on and struck by lightning. Littered and vandalised by insufficiently socialised passengers. Dented by encounters with incautious kangaroos. Parked in aging stabling yards all over the metropolitan area.

Could you really create a bankable PPP out of that sort of mishmash? This was a unique problem.

Only not.

I was old enough to remember the early days of mortgage securitisations in the eighties when I was a lawyer in London. It had seemed extraordinary that Saloman Brothers could create a saleable bond out of a collection of domestic mortgages. Issuing fixed income securities off the back of individual mortgages meant finding a way to cope with all the stuff that happens to individual mortgagors: they fall into arrears, burn down their houses, die, move interstate, make extra payments, make the wrong payments, sublet the premises in breach of contract, try to hand back the keys and generally make nuisances of themselves.

But the extraordinary becomes ordinary. By 2004, if you had any kind of an income stream, the lawyers would have a contract in their precedent library to securitise it.

The train PPP essentially posed the same problem. How to ensure the availability payments the contractor would be using to pay off the debt finance could (subject to competent performance by the contractor) remain stable and predictable notwithstanding all the out-of-the-contractor's-control stuff that happens on a crowded passenger network.

It was unique in that it hadn't been done on a railway, or so we thought at the time. But if it could be done for domestic mortgages, surely it could be done for trains?

It could. We did it.

PART 6

Collocation

Collocation is a great tool for collaboration. It isn't enough.

If you have two groups of people that need to collaborate, putting them together usually helps, particularly if they share kitchen facilities. Casual conversations at the coffee point can do a lot to oil the wheels.

But if you put those two groups of people on the same floor of a building, then fence them into two completely separate areas with separate security access so they can't have a face-to-face conversation without making an appointment, they probably won't talk any more than they did when they were separated by half a city.

It's not just a matter of physical location, it's about attitude.

Is the door open or closed?

Making the President's day

Project managers are often asked to make a special effort to give politicians a media opportunity, but this one probably takes the all-time record.

It happened on Friday, 10 October 1913 as part of the construction of the Panama Canal. President Wilson got to blow up the Gamboa Dyke.

In an early example of working from home, he did it by pressing a button at the White House.

The Reuter's report shows just how much trouble the project team went to in order to set up the link:

"Elaborate preparations had been made by the Western Union and Central South American Telegraph Companies for the instantaneous and automatic transmission of the President's signal from Washington to Galveston, a distance of 1556 miles. At Galveston it was taken up by sensitive transmitting instruments, and was sent over the cable (793 miles) to Coatzacoalcos, thence for 188 miles over the wires of the Tehuantepec National Railway, for 766 miles over the Pacific cable to San Juan del Sur, and for 718 more miles by cable to Panama. Thence it proceeded to its destination over the wires of the Panama Railroad Company."

Once the President pressed the button, it took just under 4 seconds for the signal to close the 4000-mile circuit and ignite immense charges of dynamite to blow up the Gamboa Dyke, the last obstruction to the navigation of the greater part of the Canal by light-draught vessels.

A splendid bit of PR. And I bet it made the President's day.

Get the best team

Ellesse Andrews of New Zealand is the Olympic women's cycling champion.

Her bicycle is a lot better than mine.

Which doesn't butter a single parsnip. If we swapped bicycles, she would still beat me by almost as many metres as the length of the course.

Okay, nobody can ride a bicycle with square wheels.

But if the kit is basically fit for purpose, you'll win more medals by putting your efforts into getting the best team than into getting the best kit.

If you have to ask, you already know.

I once met someone who worked on a helpline for solicitors at the Law Society in England. Usually the callers were sole practitioners, who didn't have a colleague they could consult with.

The most common problem was conflicts. Solicitors are generally not allowed to accept a client on a matter where they have a conflict of interest. Some conflicts are blindingly obvious: no, you can't act for a client who wants to sue someone if you are also acting for the party being sued. Other situations are less straightforward, which is why the Law Society offered the helpline.

The thing that struck this guy was that the answer to the question, "Do I have a conflict of interest?" was always "Yes". Every time.

It can be really hard for a sole practitioner to turn away a client. Everyone has an image of a lawyer as a corporate hotshot in a huge firm earning shedloads of money. But most legal firms are sole practitioners. They generally run a small business on the high street, with all the attendant how-to-make-the-rent cash-flow worries that go with it. Turning away a potential client is a big deal.

The call to the helpline is a small, desperate hope that maybe there isn't really a conflicts issue: perhaps they could accept the client after all and earn a much-needed fee. But if they were worried enough to call the helpline, they were always right to be worried.

So, next time something happens and it occurs to you to ask whether you have an ethics issue?

Yeah. You do.

Should you evaluate mandatory criteria first?

I'm not talking about the standard compliance check: making sure bidders either complied with the submission requirements or any non-compliances are so minor as to make it unfair to exclude them on that basis.

I'm talking about substantive pass/fail criteria, such as requiring a parent company guarantee or possessing some qualification. I prefer to avoid these altogether – megaprojects are so complex there is rarely any single factor so overwhelming as to rule out a bid entirely if it is the only thing wrong.

But if you do have pass/fail criteria, should you evaluate them first before letting your evaluators loose on the rest of the bid?

Why should you spend time evaluating a bid until you are sure the hurdle has been cleared?

I don't like this approach.

It is true that treating a mandatory criterion as a threshold question could have saved a lot of trouble in the Scottish ferry procurement fiasco (two ferries ordered in 2015 for delivery in 2018, seven years late and counting, way over budget – see Mandatory criteria above). Ferguson Marine, the successful bidder, would in theory never have seen its bid go through to evaluation, never mind winning the contract, because it was unable to provide the required guarantee.

But then if you are going to do what the Scottish Ministers did, and ignore a requirement described as mandatory, it doesn't really matter if you evaluate the requirement at all.

The evaluation process as a whole will certainly take weeks and possibly months, which is quite long enough as it is. If you don't allow the wider evaluation process to commence until after any mandatory criteria have been evaluated, you delay the procurement process as a whole.

True, if a bidder has failed a mandatory criterion, you will have wasted some time evaluating the rest of their bid.

But really, Ferguson Marine aside, how many megaproject bidders will spend millions putting together a bid but nonetheless miss ticking the critical box?

Might as well just get on with it.

The dangers of pleasing the client

The Melbourne CityLink tollway, which opened in late 2000, included about five kilometres of elevated roadway supported by columns.

Choices for construction of the roadway included the use of either T-beams or match-cast concrete. Using T-beams, it would be possible to work all the way along the five kilometre stretch of elevated road at the same time. Using match-cast concrete, construction would have to start at one end and work along to the other in precise sequence, with each concrete section fitted to the previous one.

The choice of method was left to the contractor, but apparently the client informally expressed a preference for the superior aesthetics of match-cast concrete, and the contractor decided to please the client.

One of the sections broke during construction and the entire project was extensively delayed while a replacement section was cast and cured.

Probably it does look better. But it cost a lot more.

Four-box model for professional integrity

Do we make professional integrity more difficult than it needs to be? Can companies make it easier for their employees?

There are two elements of professional integrity. The first is that you have high professional standards. The second is that you are prepared to uphold those standards in the face of considerable personal cost, such as the loss of a client.

So how do you help people move along these dimensions in the right direction? Obviously it depends where they start from:

- High standards, high acceptance of personal cost. These are the people that others admire for their integrity. Recruit them. Promote them. And, since the business bottom line remains important, consider giving them some extra training in inter-personal skills. While integrity demands you be prepared to lose a client, many clients with deficient standards can instead be educated to the point where sacking them is no longer necessary.
- High standards, low acceptance of personal cost. These are people who know when wrong is being done but are afraid speaking up will cost them their jobs. Make it easy for them. Create a culture where messengers are rewarded rather than shot.
- Low standards, high acceptance of personal cost. These people are willing to do the right thing, but for whatever reason don't know what the right thing is. Educate them. Build high standards into your ordinary business processes.
- Low standards, low acceptance of personal cost. These are the people for whom cash registers were invented – instead of a drawer full of money that is easy to steal, a cash register locks the till between transactions. When they were introduced into retail businesses, employee theft of money vanished overnight (theft of stock was still a problem, but hey.) While such people are not the most desirable employees, if your business processes make it impossible *not* to do the right thing, their integrity may never need to be tested.

Put together, that looked alarmingly like a management consultant's four-box model, so I went ahead and created one. (See *Figure 3: Four-Box Model of Professional Integrity*)

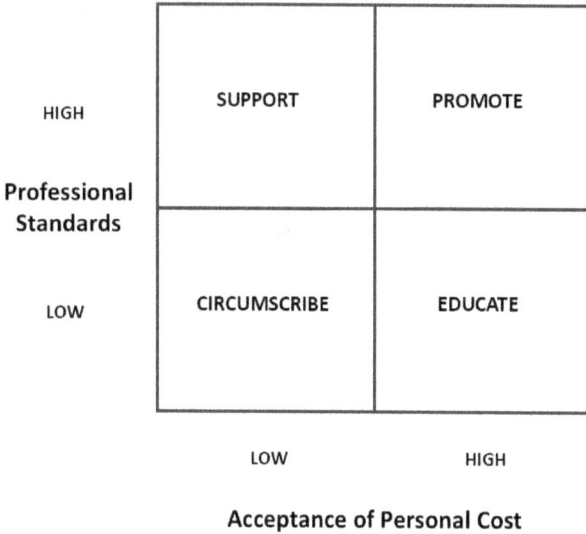

	LOW	HIGH
HIGH Professional Standards	SUPPORT	PROMOTE
LOW	CIRCUMSCRIBE	EDUCATE

Acceptance of Personal Cost

Figure 3: Four-Box Model of Professional Integrity

There you go. Professional integrity made easy.

Being unreasonable sends a message

If the other side appears totally unreasonable in a negotiation, either you've got something badly wrong or they aren't people you should be doing business with. Or both.

I have grim memories of a negotiation in the late 90s, grinding through a PPP contract pointing out problem after problem and getting zero movement from the other side.

At one point, after half an hour of driving the other side's lawyer into a corner, fighting every inch of the way, he finally admitted their performance regime formula involved double counting.

My momentary elation at having at last achieved a concession was instantly crushed, when he followed the grudging acknowledgement of double counting with the words, "But we want it anyway." (!!!)

Two days later, the other side re-issued the contract. They had at least been listening when I pointed out the problems it gave us. The clauses I had targeted had almost all been made even worse.

As clear a statement of f*** off as if they'd said the words.

We walked away.

Truth was, the other side was right – we should never have walked in. But at the time it was all caught up in politics and they could neither refuse to talk to us nor say out loud what they really thought.

We got the message.

What are the odds?

In 2016 Ireland awarded a contract for construction of a new children's hospital in Dublin. The Irish Times reported the then Minister for Health, Leo Varadkar, as saying "short of an asteroid hitting the planet", the new hospital would be built by 2020.

Since then, the date for completion has been pushed back about a dozen times, with the current earliest date being June 2025.

Did an asteroid hit the planet?

Actually, yes.

According to the University of Arizona's Catalina Sky Survey, small asteroids hit the earth several times a year, so the number of times the completion date has been pushed back is roughly equivalent to the number of asteroids that have hit the earth in the same period.

No misrepresentation there, then.

Well done, that Minister.

Giving good instructions

Why does it never occur to people who whinge about the quality of the advice they get that the failure might be in the quality of the instructions?

You may feel you need advice on the law relating to endangered species, for instance. But if you neglect to mention you need it because a site worker has found a colony of Green and Golden Bell Frogs (the critter that caused the Sydney Olympics project to re-locate the tennis centre) and you don't know whether it's still okay to send in the bulldozers on Friday, don't blame the lawyer if the answer is unhelpfully broad and doesn't arrive until Monday.

Mind you, any halfway-competent lawyer faced with the question, "What is the law on endangered species?" ought to respond, "Why do you ask?" Law schools don't attempt to teach how to tease meaningful instructions out of an incoherent client, which is a pity. Given the number of incoherent clients out there, it's a core skill.

But it would be nice if more clients mastered the art of giving good instructions.

Integrating evaluation scores

One of the difficult decisions in structuring an evaluation methodology is how to integrate scores across different criteria.

Technical matters are normally scored individually, often out of 10, and then the individual scores are aggregated and weighted to produce a total technical mark. But costs are measured in currency.

How to put these together (leaving aside the markup of the contract, which carries its own special evaluation difficulties) is always an issue. Should costs should be converted into a score, and if so, how?

In one sense, the conversion happens whether you do it formally or not. In deciding whether a cost of $3.2 billion plus a technical score of 72 is better than a cost of $3.9 billion plus a technical score of 82, you are deciding the dollar value of the 10-point differential or the points value of the $700 million differential, even if that is not how you phrase the decision inside your head.

The question is whether you should try to set those conversion values in advance of receiving the bids. I prefer not to, because a simple conversion scale is probably not going to handle the complexity of megaprojects very well.

The Nuclear Decommissioning Authority tried to deal with the issue on the Magnox decommissioning contract back in 2014 by introducing a complicated conversion scale instead of a simple one. They thought they might reasonably expect bidders to offer a range of 10 to 20% savings against the pre-tender estimate, but that savings above 35% would not be credible. The conversion scale they created was in the form of an S curve. The steepest part of the curve was set between 10% and 20% savings, to provide the greatest differentiation between bidders, with the curve going almost flat at the 35% level.

They decided probity required them to provide the S curve to the bidders. I think that was arguable, but anyway, they did it. Lo and behold, the bids came in around the point of the curve where maximum marks were available. The flat bit. That meant a difference of almost £300 million resulted in a scoring differential of less than 1.5%. Had that same difference appeared in the steep part of the curve, the difference in scores would have been much greater.

The use of an S curve was a valid choice. The losing bidder did challenge the contract award decision successfully, but the problem lay elsewhere. (See *It's an ill wind ...* above.) This approach to integrating the scores was not improper or unlawful.

It just didn't work the way they were expecting it to.

It can be a bit scary to take the approach that the proper conversion rate is not set in advance but is something the evaluation committee has to make a call on after taking everything into account. To survive challenge, it has to be a call backed with solid reasons, clearly expressed.

That's harder than plugging a number into a formula.

But it works better.

Nice work if you can get it

The Holyrood project for the construction of a new Scottish Parliament building was famously delayed and over budget. One lesser-known impact was the effect it had on advisory fees.

The UK National Audit Office (NAO) report in 2000 found that fees for project consultants had risen from £10 million to £26 million, largely because the fees were calculated on a percentage of construction costs, and construction costs had more than doubled.

The increase in construction costs was not foreseen, but the NAO pointed out the risks would have been shared more fairly if a mechanism had been included to reduce the fees as a proportion of construction cost as the level of cost increased.

Who listens to the NAO?

No action was taken to renegotiate the fees with the Holyrood consultants until 2003, which was unfortunate, as construction costs went up in the meantime a further 220 per cent. By the time of the NAO's next report in 2004, the fees had risen to £50 million.

It would be nice to think that the value of the consultants increased with the price, but somehow I doubt it.

Death by a thousand cuts

The trouble with salami is that one slice is neither here nor there, but if you keep slicing, you eventually run out of sausage.

When I was involved in the privatisation of Railtrack back in 1996, the investment bankers insisted on the need to have at least six months live testing of the on-time-running performance regimes being negotiated with the train companies, to make sure their impacts were understood.

Well, perhaps three months. We insist. Well, even two months, really. Well, so long as they're all in place before the flotation, that's what counts.

Yeah, right.

After the flotation, it turned out the performance regimes had more than a few glitches. In some cases, if a train company took steps to minimise a delay caused by someone else, it would be worse off (the classic perverse incentive), because it both bore the cost of minimising the delay and reduced the amount of the compensation payments it would receive.

The intention was to reduce delays on the railway, but the complexity of administering the regimes meant an embarrassing number of rail employees spent their days arguing about responsibility for delay instead of dealing with the causes of it.

It would have been a lot easier to fix if the problems had been visible before the parties left Government control. But the investment bankers ran out of sausage.

Good answer, wrong question

"What can I do about linens?"

The question took me by surprise and I had no idea what the student was on about.

I was presenting a lecture on procurement to a group from Bangladesh, who had come to Australia for a special programme run by Macquarie University to upskill public sector employees from developing countries.

The student explained he was responsible for procurement at hospitals, and the costs of getting linen laundered had been going up and up. His question triggered an interesting discussion on where to draw the line between what should be done in-house and what should be outsourced, and the effect this might have on value for money. What if laundry was sorted by hospital staff before it was collected by the contractor? What if the hospital owned the washing machines?

A few days later I was visiting my doctor for a check-up, and noticed that at the head of the examination couch was a huge roll of soft paper. The doctor pulled a length to cover the couch before I lay on it, and at the end of the consultation, ripped it off and put it in the bin. The high cost of laundering linen had been solved by not using linen at all.

Sometimes the right question is not "How should I best procure this?" but "Am I procuring the right thing?"

A chequered history

It's a problem any asset manager will recognise. Nobody paid enough attention to how a rail funding decision might affect what happens on the ground.

Oliver Roeder has written in the *Weekend FT* about the present problems created by a decision in the nineteenth century to divide a large chunk of western Wyoming into a chequerboard of squares one mile on each side, with squares being alternately public and private land. The private land was given to railroad companies, which sold their squares to fund the construction of the link between the Midwest and California.

Amazingly, it is only now that the obvious practical problem has emerged: how do you get from one square of public land to another without trespassing on private property? In a place where the right to go out and shoot elk and mule deer on public land is as much taken for granted as the right to shoot trespassers, the answer is critical. A current court case is being followed with interest and will no doubt make it all the way to the Supreme Court.

The really staggering thing is that an issue which was entirely foreseeable in the nineteenth century has made it all the way to the twenty-first century without being resolved. It would have been trivially easy to deal with the access issue in the original legislation, but all anyone cared about at the time was funding a railroad.

Sorting the problem out now could prove very expensive indeed.

Practising deviance

I've been struggling to read *Power in Megaproject Decision-making* by Jessica Pooi Sun Siva and Thayaparan Gajendran. Today I was somewhat floored by the sentence: "*The megaproject decision-making environment is likened to a sort of tournament whereby victory is premised at least in part on the capacity of the project teams to practice deviance.*"

Yes, well, the academic vocabulary is a bit specialised.

In this context, what practising deviance means – or what I think it means – is that members of project teams evade or manipulate the formal governance processes in various ways. If it helps the project, it's constructive deviance. If it harms the project (has "*detrimental consequences*" in academic-speak), it's destructive deviance.

Deviant behaviour apparently includes exchanging favours, leveraging professional credibility, managing the flow of information, bypassing hierarchies and so on.

What I would call, "behaving like a normal human being".

I think this book has it back to front. It isn't that the governance processes are normal and the project teams are being deviant.

What project teams are doing is normal human behaviour. The problem is that ordinary human interaction isn't enough to deliver a megaproject, because megaprojects are by definition not on a human scale. There are too many people involved, too many different interests, too many complex interfaces, too many different types of specialist expertise.

To get the job done, you need to impose artificial governance processes to direct the ordinary human interactions into productive channels. Committees, to give the project team regular access to decision makers. Procurement rules, to keep everyone on the right side of the law. Design management processes, so everything fits together properly. Project management processes, to keep it all on track.

And no, human beings aren't going to stop being human beings just because they've signed up to an artificial governance process. They're still going to be exchanging favours, leveraging professional credibility, managing the flow of information and generally doing the things people do.

Constructive normal behaviour or destructive normal behaviour? Could be either.

The real message to take away is that individual actions matter. Even on a megaproject.

PART 7

Confidentiality run mad

The public sector version of the need to know principle: if you need to know, they don't tell you. And if you do know, you mustn't tell anyone else.

When confronted with a request under freedom of information legislation, the first reaction of many government entities is to look up the exemptions section to see if they can find an excuse for not handing over the information. Not quite what the legislators had in mind.

This obsession with confidentiality may have reached a peak in the state of Victoria in 2009, when the government let a contract for a desalination plant.

Not only did the contract include a very restrictive confidentiality clause, where every word uttered by the contractor had to be vetted by the government in advance, but the clause was incorporated into the performance regime. Breach would give rise to an abatement, that is, money would be deducted from payments owing to the contractor.

Whether the "abatement" would actually have been enforceable against the contractor or whether it would have been struck out as a penalty was never tested in court.

But the purpose of a megaproject performance regime should be to incentivise performance, not pander to the paranoia of the Victorian government.

I hope they've given up on clauses like that, but I'm not banking on it.

The first employee

The period immediately after signature of the contract sets up the relationship between the parties. Regardless of whether the form of contract is nominally collaborative, collaborative behaviour will increase everybody's chances of bringing the project in on time and on budget.

So you should be getting together, client and contractor, agreeing a set of values and a code of behaviour. No, we are not going to fire off claims and instructions without having had a conversation about it first. Yes, we commit to resolving issues promptly. All those good things that build trust and keep the project moving forward.

Stick them up on a wall and keep to the commitment when things get difficult.

Or not.

On one contract, at the first meeting after contract signature the contractor was required to produce a list of the employees they had just appointed to the project.

Top of the list? A claims manager.

First meeting, top of the list.

It proved to be a clear signal of the contractor's values and behaviours. But it wasn't quite what the client was hoping for.

Not happening

Binning projects can be just as important as approving them.

Back in my railway days, I was once engaged in a review of stabling yard projects. The expectation that one of the upgrade projects would be commenced in the next year or so had allowed an operational safety case for the yard which was clearly sub-optimal. Fine on a temporary basis, but not really a safe way of operating in the long term.

It was very quickly clear to me that not only was the upgrade project not going to happen in the next year or so, it had no realistic prospect of ever happening at all.

Easy enough to deal with. A safety review, a revision of the operating procedures, and the risk to workers in the yard was quickly brought down to an acceptable level.

But if I hadn't had occasion to conduct a special review of stabling projects, the expectation of the upgrade, and the associated sub-par operating procedures, might have continued indefinitely. Someone could have been killed.

People think of an investment process as something that decides which projects will go ahead. Which is true. But it is often just as important that the process should also make it clear which projects are NOT going ahead.

If you know a project isn't happening, you can plan accordingly. But being stuck in limbo can be a dangerous place to be.

SRO missing in action

When it comes to megaprojects, the SRO is supposed to be in charge. But what if they aren't?

The UK and Australia have slightly different definitions of an SRO, but they add up to pretty much the same thing.

The UK version: *Senior Responsible Owner*: "the individual responsible for ensuring that a project or programme of change meets its objectives and delivers the projected benefits."

The NSW version: *Senior Responsible Officer*: "the agency executive with strategic responsibility and the single point of overall accountability for a project/program."

The Edinburgh Tram project didn't get either memo. It started out not having an SRO at all. The omission was noticed by an Office of Government Commerce review in May 2006 and an SRO was nominated. Unfortunately, he considered (or at least told the subsequent Inquiry into the project debacle that he considered) his role as SRO related only to the period when the tram would be operational and not to the construction period.

Not helpful.

Round about June 2009, somebody finally noticed the absence of anyone performing SRO duties, and another SRO was appointed. But, as the Inquiry noted, the appointee was inappropriate because of his position as chairman of **tie**, the delivery agency. As the City of Edinburgh Council (CEC) was the owner of this project, it would have been better to have had an SRO from within CEC, who would among things have been motivated to address the issue of unsatisfactory reporting. (So unsatisfactory, indeed, that the Inquiry report included a recommendation to consider introducing criminal penalties for misleading project funders.)

CEC didn't appoint an SRO until 2011, when better governance arrangements were finally sorted out at roughly the time when the project was supposed to have been finished.

It took another three years to get the trams running.

When the risk transfer fails …

Transferring risk to the contractor doesn't work well if the contractor goes bust.

It happened in two of the three controversial PPPs for the modernisation of London Underground's infrastructure, awarded in 2003 to a consortium called Metronet, which set up two Special Purpose Vehicles, one for each contract.

PPP contract technology was not then as advanced as it is now, and although much risk was supposedly transferred to the contractor, there was no control over the terms of the subcontracts.

All five members of the consortium had contracts to perform work for their SPV. This was not of itself an issue. They had, after all, formed the consortium precisely because they wanted to win the work. But having won it, they awarded themselves sub-contracts that were, shall we say, not on fully arms' length terms. One contract, with the rolling stock manufacturer, was found by the PPP Arbiter to be operating efficiently and economically after a somewhat rocky start.

The real problem was with the other contracts, which were sub-contracted through another SPV, Trans4m, owned by the remaining four shareholders.

The contract with Trans4m gave Metronet very little in the way of enforcement levers – invoices had to be paid when presented, with no withholding rights for performance failures – and there was little opportunity to contract work elsewhere on more reasonable terms. The chairman of Metronet was reportedly threatened with litigation when he sought to reduce the amount of work going to the shareholders.

By 2007 their costs were, between them, some £2 billion over budget. While some could be claimed by Metronet as not having been anticipated in the bids, the independent PPP Arbiter would only award costs that would have been incurred by a contractor performing in an efficient and economic manner. An interim review indicated the amount recoverable could be less than £500 million. The balance, a billion and a half pounds, represents a staggering amount of inefficiency.

Metronet went bust. What a surprise.

So much for the risk transfer. London Underground services are somewhat critical to London's economy, so the UK government had to pick up the pieces.

In addition to the direct costs, estimated by the National Audit Office as being up to £410 million, 95 per cent of the Metronet debt finance had to be paid off by the Government immediately. Ouch. An un-programmed payment of £1.7 billion makes a dent in anyone's budget.

Treasury was not amused.

What is it about ferries?

"Ferry Fiasco" headlines have been appearing in Tasmania and New Zealand in 2024 as they join Scotland in showing the world how not to purchase ferries.

Scotland's fiasco dates back to an order for two ferries placed in 2015 for delivery in 2018. The contractor went bust and was nationalised by the Scottish government, which did not noticeably improve performance. So far only the first ferry, delivered about seven years late, is in service and the eventual total cost will be at least four times the initial estimate.

In October 2024 Tasmania lost a Treasurer, who resigned after accepting responsibility for what has been described in the media as "a monumental stuff-up": two ferries, $500 million over budget and five years late. The first ferry has been completed, but since there is no wharf ready to receive it, nor likely to be until at least 2027, it spent Christmas 2024 in Edinburgh, a temporary berth. The Tasmanian government had hopes of leasing it to Scotland, which is currently a ferry short of quota (see previous paragraph).

New Zealand was also buying a couple of inter-island ferries, intended for delivery in 2026, but decided to cancel the contract with the South Korean shipyard in early 2024, at a cost estimated at around half a billion dollars. At the time, it was suggested the ferries would be replaced by alternatives that were more suitable and cheaper. Ooh, look, there's a pig flying past. No announcement of where any future ferries might be coming from has yet been made.

People have been building ferries for centuries. Why is it suddenly so hard?

Small project, big risk

One advantage of a megaproject is that you don't have any trouble convincing people of the scale of the risk.

It is axiomatic that anything costing a billion plus is bound to contain big scary risks that need attention from the C-Suite. Conversely, a project with a small price tag is presumed to be low risk, able to be consigned to a junior project manager to cut their teeth on.

Sometimes the price tag is a poor indicator of risk.

I recall a very small signalling project from my Railtrack days in London. Feasibility work was progressing nicely under a bright but inexperienced project manager. When it went up for approval, one of the umpteen signatures required belonged to a senior signalling engineer who had been with British Rail more or less forever.

The approval papers got no further than his desk.

The junior project manager was summoned. Possibly a dead body was mentioned. In any event, he was told in no uncertain terms that, while his project would only have required replacing a couple of widgets in the relevant signal box, the equipment in the box in question was very old and very fragile and some of it would probably disintegrate if touched, wiping out train services across a sizeable percentage of West London.

The engineer was never going to allow anyone to touch the signal box, however small or cheap the project, unless they were prepared to replace the whole thing. The risk was just too great.

End of small project.

Risk by design

An engineer told me a story from the somewhat distant days of his youth, when one of his first projects had been to design a crash barrier for a bridge.

He proudly produced a design that would withstand an impact from the heaviest possible truck travelling at maximum speed. His boss looked at his handiwork, and asked a question: "If the heaviest possible truck did hit the crash barrier at maximum speed, what would happen to the bridge?"

The engineer didn't know the answer to that, so he went away and did the calculations. Oh. And indeed ah. The impact would cause the entire bridge to collapse.

Embarrassed, he redesigned the crash barrier.

Yes, in the worst case, the truck would go through the barrier into the ravine. But although it would be a seriously bad outcome for that particular truck driver, it would be a great deal better outcome for everyone else.

Net present value

Net present value is a useful concept for evaluating megaprojects, but it has its limitations.

Money is assumed to generate a return, so incurring a cost now means you lose the benefit of that return immediately. Incurring the cost later means you have the benefit of the return in the meantime. The present value of a cost to be incurred later is calculated by applying a discount rate, which varies according to the assumed rate of return. Similarly, revenue now is worth more than revenue later. The same discount rate is applied to future revenue to determine its present value.

The NPV of a project is simply the present value of the forecast revenues minus the present value of the forecast costs. A positive NPV means the project will be profitable (assuming the costs and revenue forecasts were accurate). A negative NPV means Treasury doesn't want to know.

What I have never been able to get my head around is where the asset itself fits into this. Depending on the discount rate, the revenues will have been discounted back to zero after fifty years. The Sydney Harbour Bridge has been around almost twice that length of time. Most of London's bridges are more than 100 years old, some of them a lot more. The Pons Fabricus in Rome is still in use after more than two thousand years. Is that worth nothing?

And what about the value of a transport corridor? Many Roman roads are still in use today, the alignments taken over to become part of the modern road network. After the Beeching cuts in the UK in the 1960s, some of the redundant railway lines were turned into roads, but others were built over. Much as some local councils would like to restore those lost lines, the cost has become prohibitive because the transport corridor was lost.

NPV is a valuable tool. But it doesn't measure everything.

The red button

Your governance structure is like the red button on the factory floor.

What's the red button? That's the one any worker on the line can press to stop production if there's a problem with quality. The first ones in Japan were pull cords, but probably everyone has now progressed to buttons. It's an important part of a Total Quality Management system.

Executives often see the governance structure of a megaproject as being there to give them oversight. A control function.

Which is sort of true. But the governance structure is not just there to ensure that messages from the top get all the way down. If the CEO wants to find out what is happening somewhere in the organisation, they don't need a governance structure, they can get off their arse and go and look.

The line worker, on the other hand, has no access to the CEO's office to find out whether the problem they tried to report has been noticed by someone able and willing to do something about it. If your governance structure doesn't have a way for stuff at ground level to get elevated to the C-Suite, it isn't doing its job.

But it isn't enough to have a red button. The US firms that tried to copy the Japanese TQM systems in the 1980s mostly fell a long way short. No surprises there. Workers won't press a red button unless they know what will happen. And if the answer is a) nothing or b) the messenger will get shot, the button never gets pressed.

So yes, your governance structure is like the red button on the factory floor.

Useless if nobody is willing to push it.

It's not a hardware problem

I saw a media report recently of a CEO trying to convince a town hall meeting of employees that corporate culture issues didn't lie with top management. The problem in his eyes was lower-level business managers and HR staff who failed to escalate allegations of sexual harassment and bullying.

It took me back to a high school encounter with a computer programmer. I can't at this distance recall who he was (we are talking 50 years ago), but he had programmed one of the earliest computers. So early, it worked on electric valves. Whenever a program failed to execute properly, his first task was to go round and tap all the valves to check they were still functioning, because a blown valve was the most likely explanation.

It must be comforting to a CEO to think that the problem lies down there in the electric valves not doing their job.

But these days, as any programmer knows, if your program fails to execute the chance that the problem is in the hardware is vanishingly small – the fault is in the way you wrote the program.

CEOs take note.

Spotting the difference

When I worked on the Waratah project, buying new trains for the Sydney rail network then operated by RailCorp (now Sydney Trains), we knew we were creating major change for the organisation. When you sign a billion-dollar contract, your business is about to become a billion dollars' worth of different.

Sometimes in ways you don't expect.

About a third of the electric fleet was being replaced in a relatively short timeframe. Thousands of drivers and guards would have to be trained to operate the new trains. The passenger timetable would be rewritten to accommodate the superior acceleration and braking characteristics of the new trains. The working timetable would be rewritten to accommodate different maintenance cycles and locations. There was a major restructuring of the maintenance workforce. New infrastructure to support the trains – stabling roads, a maintenance facility, substations – would have to be maintained. The old trains would have to be disposed of.

The project to change the organisation was as big as the project to manufacture the trains.

And then there was the change we didn't think of.

Well, okay, we thought of it eventually, and fortunately still in time to sort it out, but it took us a while.

The trains that were being scrapped, the old S-Sets, were the ones used to train new drivers and guards. Everyone learned first on an S-Set. When they were sufficiently experienced, they might then go on to crew other train types, such as the Tangara.

Which meant all the training materials and courses for the rest of the fleet assumed the crew already had the skills and knowledge they would have picked up from working on the S-Sets. They were conversion courses, not train-new-crew-from-scratch courses.

So how were we going to train new crew once all the S-Sets were gone?

A simple question, once you've asked it. But it took us a while to ask the question.

Unsung plus of a PPP

The great thing about a PPP is that there's no need for liqui-dated damages.

I mean, you can, but why bother?

LDs are a pain. Yes, they can be a very effective incentive on a contractor, but there are downsides.

One is that you have to demonstrate the amounts to be paid or deducted are a genuine pre-estimate of the loss suffered in the event of delay, otherwise the LDs will be unenforceable. Estima-tion is rarely straightforward and delays may even *reduce* cost, because a new asset offering new benefits can result in increased operating costs that aren't incurred if delivery of the asset is delayed.

The lawyers can often find ways around that, but one thing they can't get around is that if there is provision for LDs, there must also be a provision allowing for extensions of time where the delay is not attributable to the contractor. This is wearing to administer and is usually a fertile source of disputes.

A less obvious problem is that the enforcement of LDs has to be a decision made at the relevant time. When a megaproject contract is signed, the responsible Minister is likely to brandish the LD clause as demonstrating they have the ability to get tough with an under-performing contractor. The same Minister is usually hiding behind the door when it actually comes to enforcing a contract against a major local employer who cries poor and threatens job losses in marginal constituencies.

But a PPP? No worries. The financing plan for a PPP shows the moment of maximum loan drawdown at about the time the project is due to be delivered. If the availability payments don't kick in because the contractor is late, the horrendous interest bill provides a daily incentive quite as big as any LDs you could justify.

What does the Minister have to do? Absolutely nothing. Always an easy ask.

Defenestrating good practice

Portcullis House, a terrorist-proof building built for the House of Commons at Westminster in the late 1990s, included what was described in the subsequent court case as "*probably the most expensive fenestration system ever*".

The tender documents for the fenestration system (i.e. the windows) contained an unfortunate omission.

No evaluation criteria.

Publishing the evaluation criteria up front should be a no-brainer. Quite apart from any consideration of fairness, how will bidders know what to offer you if they don't know what you will base your decision on?

The losing bidder had offered the lowest price and sued the Corporate Officer of the House of Commons for failing to award them the contract.

The House of Commons lost. It was unable to persuade the court the bidders could have worked out the other evaluation criteria from the information requested.

Not particularly surprising. The court also found the project team never did agree what the criteria were. If even the project team didn't know, how could the poor bloody bidder work it out?

Corporate conflict

Conflicts of interest don't just happen to individuals, they happen to companies – and it can be very bad news for a project. Witness the Edinburgh Tram project: three years late and more than 40% over budget to deliver a line ending two kilometres short of the originally intended destination.

The conflicted organisation in this case was **tie**, which was incorporated in 2002 as a company wholly owned by the City of Edinburgh Council (CEC) to manage the planned scheme for congestion charging and to use the funds from that scheme to deliver various transportation projects, including a tram network for the city.

Things didn't quite work out that way. Congestion charging was abandoned, as was the Edinburgh Airport Rail Link, and the Scottish Ministers withdrew **tie** from its management functions in the Stirling-Alloa-Kincardine railway. By late summer 2007 the tram project was the only one left in **tie**'s portfolio. If that didn't go ahead either, **tie** was doomed.

CEC's decision to proceed with the project relied entirely on **tie**'s cost forecasts and risk assessments. Since the very existence of **tie** and the jobs of everyone in it were dependent on the project going ahead, it would not be surprising if these were a tad optimistic.

In the event, they were so blatantly and unjustifiably optimistic that the subsequent Commission of Inquiry included recommendations not only asking the Scottish Ministers to consider civil proceedings against individuals providing misleading information to local authorities, but also to consider introducing a statutory criminal offence.

Given the in-hindsight-obvious conflict, CEC should have sought independent review of the cost and risk information they received.

But they didn't.

Understanding the objectives

If you don't know why, you'll mess up the how.

There was once a UK company which was keen to avoid tax. The accountant suggested operating through a Cayman Islands company. Only problem was, meetings, including the Annual General Meeting, had to be held in the Cayman Islands, otherwise the tax avoidance scheme would be rendered void.

As far as the chairman was concerned, this wasn't a bug, it was a feature. He adored scuba diving. So every year the directors would toddle off to the Caribbean for sun and scuba diving and hold the AGM somewhere in between martinis.

And then the chairman retired.

By that time, as far as anyone in the company realised, the chairman's love of scuba diving was the only reason for holding the AGM in the Caymans.

So when the incoming chairman declared a preference for golf, they held the AGM at St Andrews.

Bad move.

PART 8

Paying bid costs

The real issue about bid costs in megaprojects is not whether the client should pay them or not, it's why they're so high in the first place.

Regrettably, I think a lot of it is down to client behaviour. Some poor but sadly not uncommon practices:

- Changing the timetable or requirements without notice, so bidders incur wasted preparation costs.
- Requiring unnecessary information, so bidders pay to provide stuff that doesn't affect the procurement decision.
- Shortlisting too many bidders, putting up the overall cost.
- Taking forever to make decisions and respond to queries, so expensive bid teams are kept hanging around.

- Putting forward indefensible contract conditions or risk
  allocation, so bidders incur legal costs objecting to
  something that should never have been proposed in the
  first place.

Even if the client doesn't pay bid costs up front, they end up
paying sooner or later, because higher costs mean higher prices.

Why make those costs higher than necessary?

Shovel-ready projects

When it comes to transport projects, what's the difference
between the emperor Nero and John "Two Jags" Prescott?

Nero used a pickaxe.

John Prescott, then UK Deputy Prime Minister, cut the first sod
in the construction of the high speed rail link from London to
the Channel Tunnel in 1999 using a spade, which was specially
engraved and presented to him.

The emperor Nero also kicked off construction of a new transport
link, in his case the Corinth Canal, but he used a pickaxe. The first
bucketful of rubble was filled in AD67. History does not record
whether the pickaxe was specially engraved and presented to the
emperor afterwards, but it's the sort of thing Nero would have liked.

Prescott's project did rather better than Nero's. The first leg of
the Channel Tunnel Rail Link was completed in 2003, just 4
years after the start of construction. The Corinth Canal was
finally completed in 1893, more than 1,800 years after Nero
wielded the pickaxe.

And you thought your own project was running late.

Process safety

The 2005 Texas oil refinery disaster (15 dead, 170+ injured) provides an object lesson on the difference between personal safety and process safety – and why taking care of personal safety is not enough.

The subsequent Baker report gave a neat definition of the two types of hazard, quoted in Andrew Hopkins' excellent book *Failure to Learn: The BP Texas City Refinery Disaster*:

"Personal or *occupational* safety hazards give rise to incidents—such as slips, falls, and vehicle accidents—that primarily affect one individual worker for each occurrence. *Process* safety hazards can give rise to major accidents involving the release of potentially dangerous materials, the release of energy (such as fires and explosions), or both."

The proximate cause of the disaster was the overfilling of a distillation column, which allowed a geyser-like eruption of liquid petroleum from the top of the column, creating a vapour cloud that ignited in a massive explosion. There was a whole series of process safety issues, both in how the overfilling was allowed to occur, and why the consequences were so serious when it did.

The tragic irony is that there had been a meeting in the control room half an hour beforehand. The plant management wanted to celebrate. A 35-day maintenance shutdown for two other process units on site had just been completed without a single recordable injury and with only two first aid treatments. A genuine occupational safety achievement. Well done.

Ten minutes after the meeting broke up, boom. Fifteen dead.

Doing the reference check

Failure to check references is very poor practice. Ignoring negative references borders on the suicidal.

A major issue in the Queensland Health payroll debacle back in 2010 was the highly damaging role played by an individual contractor to CorpTech, a division of Treasury that was managing the project on behalf of Queensland Health.

He was originally engaged, without any competitive process, to conduct a review of the project. His review recommended the creation of a powerful position to which he then contrived to be appointed, again without any competitive process.

Astonishingly, the appointment was made notwithstanding that the firm which had previously engaged him explicitly refused, in writing, to provide any recommendation.

The contractor proceeded to exhibit, in the words of the subsequent Commission of Inquiry report, *"exuberant indiscretion"*. He encouraged the adoption of a "prime contractor" model for the project and assumed de facto control of the evaluation process to select the prime contractor. He then intervened improperly, triggering a reassessment of scores that enabled the contract to be awarded to a tenderer who should not have been appointed.

The failure of the subsequent implementation of the new payroll system was described by the subsequent Commission of Inquiry as *"catastrophic"*.

There were other factors, of course. One individual, however exuberant their indiscretion, is unlikely to be the sole cause of a catastrophe. But ignoring a negative reference is never going to be a good move.

Dispute timing is never convenient

Megaproject disputes happen at the most inconvenient time. That's not an example of Sod's Law, it's the way it works.

While construction is still in progress, the parties are usually hopeful of a negotiated settlement and both are desperate to get the construction completed. The client because the need for the infrastructure is more and more pressing, and the contractor because they want to stop spending money they are no longer sure they can recover. It's not until the infrastructure is up and running that the writs come in.

When they do, the lawyers turn out in force and start amassing the evidence to support or defend the claims.

But where is the evidence? Yes, there's a mountain of paperwork, but key information is always inside the heads of the project team. They have to spend days with the lawyers working up their statements, and then they are needed to give evidence in the arbitration.

They're needed – but they're not there.

When construction is complete, project teams vanish. Certainly from the project, often from the company, sometimes from the country. They probably won't give evidence at all without being offered an indemnity, and they have no interest in re-living old disputes. They've moved on. There's no one to interpret the files. Decisions about the conduct of the arbitration have to be made by people who had little or no involvement with the project in the first place.

Anyone who has suffered a big arbitration – and I do mean suffered – will tell you it's a huge time and money sink. Costs of

$1 million a month for each party involved are by no means unheard of, and they can go on for years. And years.

It's undoubtedly no accident the impetus to adopt Dispute Avoidance Boards – which have a splendid record in preventing disputes on major projects – tends to appear immediately after major arbitration or litigation. "We're never going through *THAT* again," is an entirely rational response.

It's just a shame so many people have to go through a major dispute before they understand the value of avoiding one.

A business case

It may not always feel like it, but a business case isn't a form you fill in, it's a compelling reason for doing a project.

A NSW Treasury presenter once fielded a complaint about how Treasury kept knocking back projects and taking forever to do it. (Probably a regular occurrence, come to think of it, but the incident I'm thinking of was about ten or fifteen years ago.)

The presenter's response was essentially, "No, we can come up with the money quite quickly if you come up with the business case." She pointed at me and gave one of my projects as an example.

I was amused, because for that particular project, I hadn't submitted a formal business case at all.

I had been working on a fleet strategy for the Sydney metropolitan railway, figuring out how many trains would be needed and where and how they would be stabled and maintained. I was advised very late in the piece that the agency building an extension to the railway was about to go out to tender for a stabling yard. The extension team hadn't been involved in the strategy development and their business case only covered about half the number of stabling roads we now knew were going to be needed within five years.

So I basically ran round to Treasury and said please please please let the agency include an option in the tender to build the other half of the stabling yard, because it will be about $50 million cheaper if we do the whole lot in one go.

And Treasury, which is always pleased to save $50 million, said yes.

Had I filled out a business case form? No.

Did I have a damn good business case? Yes.

And that's what counts.

To escalate or not to escalate

We'll all make a push to discourage a client from doing something stupid. It's a bit more difficult if they've already done it.

In the InterCity West Coast franchise case, back in 2012, the UK Department for Transport advised bidders of the amount of the subordinated loan facility each would have to provide if successful in the tender for the £5.5 billion franchise contract. The amounts were calculated using a process that didn't match advice previously given to the bidders.

The external lawyers hadn't been represented at the meeting where the calculations were done. The lead partner found out about it after the amounts had already been notified to the bidders and did raise a concern both with the project team leader and separately with the internal lawyer. The concern appears to have been expressed in relatively low-key terms and no one escalated the matter within the Department.

The external lawyers formed the view that the issue was closed and it would be "*unhelpful*" to provide further advice.

Bad call. The losing bidder sued and the whole procurement process had to be cancelled.

Lawyers make a lot of money advising governments and wish to be seen as helpers rather than troublemakers. They are acutely aware that their advice is sometimes inconvenient and also that advisers whose advice is too inconvenient may not be welcome back.

The task for the client is to build a relationship with the lawyers that encourages them to stick their hand up when you have made a complete bog-up of something.

Better to hear it from them now than read about it in the statement of claim later.

You know what?

I've never quite understood why people feel so threatened by the expression "It's not what you know, it's who you know."

This is not some manifestation of malignancy, it's normal social behaviour. A decision isn't made by a what, it's made by a who.

Okay, getting airtime with the who can certainly be an issue, particularly for those who start from a position of disadvantage. There are gatekeepers at the who's office door, and the gatekeepers may not believe you have the what. Largely because, in their somewhat jaundiced experience, the world is full of whatless individuals with delusions of relevance.

If you have enough what, the who are really quite keen to make your acquaintance.

When you think about it, the entire governance structure of a megaproject is designed to that end.

The committees and consultations plug every who into the process, and the meetings, the project reports, the contract early warning notices and all the other manifestations of governance are there to see that the who give proper consideration to the what.

Doesn't always work perfectly, of course. But that's what the governance system is there for: to make sure the what gets introduced to the who.

Rule for a successful IT project

Practical completion is not the day the new system goes live. It's the day the old system goes dead.

Let's face it, you don't normally invest in a new IT system to start a new business. You do it to make the existing business work better. You won't get the productivity and quality benefits of new business software if people go on using the old business software alongside it – or worse still, instead of it.

Top management is rarely interested in what happens after practical completion. Once a new system is up and running, they mark the job as done and focus elsewhere, leaving a depleted project team to tidy up the loose ends.

Turning off the old system isn't a loose end. It's the whole point of the exercise. If management focus moves on before that's done, the job isn't finished. A long-suffering IT department ends up supporting legacy software until it's not so much "legacy" as "heritage". Business systems won't die unless you kill them.

Don't structure an IT project around what you have to do to turn the new system on. Structure it around what you have to do to turn the old system off.

Open for business?

A Business Day isn't always what you think it is.

In Australia, Christmas comes in summer. The entire country is more or less closed for non-tourist business from Christmas Eve to New Years' Day and the school holidays continue until after the Australia Day public holiday on 26 January.

Cunning contractors have been known to deliver large quantities of documents or product just before 5pm on Christmas Eve, hoping the review period specified in the contract will have expired before anyone notices.

To prevent this happening on the Waratah train procurement, we re-wrote the definition of Business Day to exclude the entire period from the Monday before Christmas to the Friday after New Year, so the clock stopped ticking during the break.

It worked really well, so we included a similar provision in the next contract we issued.

Alas, the contractor completely failed to notice and was badly caught out on their scheduling through not allowing for the longer design review period.

36 hours

That's how long it took to convert thousands of miles of track in the US South from 5'0" to 4'9".

In February 1886 the various operators met and agreed that the gauge changeover would be made on 31 May and 1 June.

To make that possible, only one rail was moved. The inside spikes were hammered into the new positions in advance, ready for the final blows of the sledgehammer when the rail was moved. Close to the changeover, spikes were pulled from the rail at intervals to reduce the number needing to be shifted on the day.

A few days before May 31, the operators began clearing cars from their lines and reducing the gauge of areas of track that could be freed up.

Some of the rolling stock was prepared in advance by machining the axles to the new gauge and placing a special ring inside the wheel to hold it to the wider gauge until the changeover. On the day, the wheel was taken off, the ring removed and the wheel replaced.

As with the track, some of the rolling stock was able to be converted in the days leading up to the changeover. But there was still an enormous amount to do. Lathes and crews were stationed at various points throughout the South to convert the rolling stock during the final push on the track.

The whole thing went off with almost no disruption to rail traffic.

Amazing what you can accomplish in 36 hours.

Note for railway buffs: the reference to 4'9" was not a typo. They went with 4'9" because it was the gauge of the busy Pennsylvania railroad. Standard gauge rolling stock (4'8½") could still manage on the slightly wider track. The final conversion to standard gauge was accomplished some years later during programmed maintenance.

Delivering with the B-Team

Bid with the A-Team, deliver with the B-Team. It's a pain for the client.

But it can be just as much of a pain for the contractor when government turns up with its own B-Team after contracts have been signed.

Why don't bidders demand key personnel restrictions?

Government will almost always try to prevent the contractor getting away with substandard performance by putting key personnel restrictions into contracts. It never works perfectly, but requiring comparable levels of qualifications and experience does give at least some influence over the selection of replacements when key personnel leave the project.

Contractors never demand the same of government. Yet the government's contract management team can make a huge contribution to the success or failure of the project. Often it's as much about managing government stakeholders as it is about managing the contract.

Probably government wouldn't agree to such a provision. Sadly, one of the reasons they might not agree is that they sometimes don't get around to thinking about a team to manage the contract until after it's been signed, so there are no key personnel to try to bind to the project anyway, possibly not even job descriptions with minimum levels of qualifications and experience.

Bad move. The quality of the contract management team matters.

Same difference

There is a disturbing parallel between Australia's Inland Rail project and California's high speed rail project.

Neither has figured out whether they can do the difficult bits.

California is trying to create a high speed rail link between Sacramento and San Francisco in the north and Los Angeles and San Diego in the south. There are two main engineering problems not yet solved: how to get through the Tehachapi Mountains at the Los Angeles end and how to get through the Pacheco Pass at the San Francisco end.

It has been estimated that getting through those two bottlenecks will make up nearly 80% of the total estimated cost of completing the project. (That's 80% of a number that has gone from $33 billion to $128 billion since 2008.)

Someday, they may figure out how to do this. In the meantime, construction is proceeding on the middle bit, from Bakersfield in the south to Merced in the north.

No, I hadn't heard of them either.

The expected cost for that section is $35 billion. Yep, more than the total original cost. For delivering a rail link that is not only a lot shorter but is also pretty much useless without the northern and southern connections. Yay team.

Australia's Inland Rail project has a similar problem, although happily not on the same scale. It's building a rail freight link between Brisbane and Melbourne. But it got started on the middle bit without knowing where it was going to put the terminals at the northern and southern ends.

This is largely a planning rather than an engineering problem, but the engineers can't come up with credible cost forecasts if they don't know where the terminals will be.

The estimate for the Inland Rail project, $31 billion (up from $16.4 billion), is therefore distinctly dodgy. As Kerry Schott stated in her 2023 report, *"In my view this cost estimate should not be accepted by the Shareholder as there is insufficient certainty about the scope, the related schedule, and delivery costs to have any confidence in the numbers."*

Quite. Any project looks better if you leave out the difficult bits.

Reorganising the railways

In 1996 the then fully-integrated State Rail Authority of NSW gave away its freight business. It then spawned the Rail Access Corporation, to own the track, and, to maintain the track, the Railway Services Authority.

The Railway Services Authority morphed into Rail Services Australia, then in 2000 merged with the Rail Access Corporation to form the Rail Infrastructure Corporation.

The Rail Infrastructure Corporation gave away the country network to the Australian Rail Track Corporation and on 1 January 2004 rejoined the State Rail Authority to form Rail Corporation NSW (RailCorp).

This exciting and ultimately pointless journey took in the Glenbrook and Waterfall accidents and saw four SRA Chairmen and six SRA Chief Executives consigned to the dustbin of railway history within a seven year period.

RailCorp itself was binned in 2014 in favour of a new split between Sydney Trains (fully integrated in the metropolitan area) and NSW Trains (a regional train operator).

It's ten years since the last major restructure, so if history is anything to go by, we should be about to embark on another round of taking the railway apart and sticking it back together again.

Let's not.

Procurement code of practice

Should megaprojects comply with the agency procurement code? Quite possibly not.

Many government agencies have in recent years devoted a lot of effort into putting really good procurement processes in place and mandating their use. Given that the agencies are letting hundreds and in some cases thousands of contracts every year, there are enormous benefits associated with having a clear and consistent process that minimises organisational friction while ensuring all the probity boxes get ticked and all bidders are fairly treated.

The one-size fits-all approach is perfectly reasonable: so what if a few contracts could have delivered, say, a 10% better outcome with a different process? That is well and truly offset by the benefits of consistency across the organisation.

Problem. Megaprojects don't do consistency. They're a one-off, pretty much by definition. And a 10% better outcome on a billion-dollar project delivers 100 million dollars of cash or value.

Look harder at the detail: is that really good procurement process entirely fit for a megaproject?

Have to have three quotes? Sometimes you need them but you probably get better pricing on a megaproject with two keen bidders going head-to-head.

Bidder must be a limited company? The consortium may not set one up until the contract is awarded. And maybe you should consider permitting trusts.

Mandatory disqualification for non-compliance with submission requirements? Better allow yourself some discretion rather than be forced to disqualify a bid (which probably cost millions to produce) because server problems meant the documents didn't get through until half an hour after the deadline.

Mm. Time to read that procurement code you claim to be complying with and make sure it really is fit for a megaproject.

Tadgell's Bluebell

I have no idea who Tadgell was, but his bluebell was a right nuisance.

We were about to seek approval for the preferred bidder decision in the Waratah train procurement back in 2006, when the Department of the Environment issued a circular about sightings of endangered species.

Oh dear. One of said species, Tadgell's Bluebell, had been spotted sometime between about 1872 and 2006 (the date was unhelpfully not recorded) on the designated site for the maintenance facility. It wasn't so long ago that the Sydney Olympics had had to re-locate the tennis courts because of the discovery of a colony of Green and Golden Bell Frogs.

Okay, some unnamed busybody botanist had seen the thing once in more than a hundred years. Maybe we could just ignore it and hope for the best? Alas, not a really acceptable approach on a $3.6 billion project. There was a clear risk that additional works would be required to relocate either the facility or the bluebells.

We could have stuck an extra clause in the contract requiring the contractor to sort it out. Unfortunately, some hasty botanical research established that not only is Tadgell's Bluebell indistinguishable from any other bluebell except when in flower, the flowering season was already upon us. If we left the matter to the contractor, the project would be blighted by the uncertainty for almost a whole year until the next flowering season. Not an option.

So we arranged for a couple of botanists to go on a course – that's right, your average botanist can't tell bluebell species apart either, even if they are in flower – and then sent them off to inspect the site.

Which took them a while, because the entire post-industrial wasteland selected for the maintenance facility was a positive meadow of bluebells. If Instagram had been invented people would have been queueing up for a selfie.

Fortunately, the chief threat to the continued existence of Tadgell's Bluebell is not habitat destruction but other bluebells.

The more abundant the common or garden species of bluebell, the more likely that Tadgell's Bluebell will be genetically swamped (which sounds unpleasant, but is reportedly painless).

And that seems to be exactly what had happened. Lots and lots of bluebells, and not a single Tadgell in sight.

Problem solved.

Postscript: After writing this, I decided it was time I asked Professor Google who Tadgell was. His name was Alfred James Tadgell, born 1863, died 1949, and he was one of the early members of the Field Naturalist Club of Victoria.

Tadgell was an accountant, but in his spare time (of which he must have had a lot) collected thousands of Australian plants and corresponded with leading botanists in Australia, England and America. His collection and most of his botanical literature is now housed at the National Herbarium in Melbourne.

Now you know.

PART 9

Financial complexity

How complicated can a finance deal get?

In 1987 Alan Bond, not yet in jail, bought the Channel 9 media group from Kerry Packer for $1.056 billion, at the time the biggest deal ever done in Australia. It was not one of the most astute deals Bond ever did – Packer later bought it back for a fraction of the original price – but I was a very junior lawyer on that deal and I still remember the financial close.

It involved one man handing over an envelope.

In the envelope – you think I'm going to tell you there was a banker's draft in the envelope, don't you? Actually there were eleven, ten for $100 million each and one for the loose change. The bank systems couldn't cope with more than nine digits on a cheque.

Just short of twenty years later in 2006, the financial close on the Waratah deal, a $3.6 billion rolling stock PPP, involved a complex protocol, several dry runs, a room full of bankers and computers and about half a day.

A bit of a contrast with the hand-over-an-envelope scenario.

Have financing deals become any simpler since then? I doubt it.

A whale of a problem

If the measurement shows you have a problem, change the measurement.

End of problem. Not.

My attention was drawn to the "London Whale" by Henrico Dolfing's recent case study on the 2012 financial debacle that left JP Morgan with a loss of about $6 billion. The study focuses on the failure of the Synthetic Credit Portfolio Value at Risk (VaR) Model, a sophisticated financial tool intended to manage the risk associated with the bank's trading strategies.

But the London Whale is also an excellent example of rules being fudged when following the rules would give an unpalatable answer. Let's face it, whales don't do governance.

JP Morgan traders took out large positions in credit default swap indices linked to the default likelihood of US corporates. The positions became loss-making, and instead of closing out or hedging the positions, the traders doubled down. That's the response of a gambler, not a risk manager, which is where the name London Whale came from – "whale" is a slang term for a gambler who makes huge bets. They hoped the market would rebound in their favour. It didn't.

Managerial and monitoring issues compounded the problem. The Head of the Chief Investment Office did not receive full position reports and aggregated risk metrics concealed the problem within the wider portfolio. The profit/loss statements for the credit default swaps that he did get were flawed, partly because of financial modelling weakness and partly because the traders were the ones choosing the values to enter in the model. (At least once, traders waited until 20:00 London time to mark their trades in the hope of getting a better price during New York trading, and subsequent investigation could not always document that the price claimed had ever been within the bid-ask spread on the day in question at all.)

Scheduled risk management meetings were poorly attended and often delayed. Always a bad sign – that applies to projects as well as trading portfolios.

But the real classic of poor governance is that the VaR limits were breached more than 300 times in early 2012. This should have rung alarm bells all over the shop. Instead, the problem was "solved" by putting a new, more lenient VaR model in place, which cut the initial loss estimates from the CDS positions by half.

Naturally, this compounded the problem, because it allowed the traders to increase the size of their bets. The positions became so large that when JP Morgan finally twigged to the extent of the problem there was insufficient liquidity in the market to allow them to be unwound quickly: the losses rose from $2 billion to $6 billion before the gaping hole in the accounts was finally plugged.

What gets measured gets managed. What gets measured badly gets managed badly.

Milestone selection

Starting to "cut steel" is a big deal in shipbuilding. It looked like a good candidate for a milestone payment on the 2015 Scottish ferries contract.

And why not? The design has been developed over many months of hard work, the drawings are now complete and construction can begin. The shipyard puts on a special ceremony and invites some VIP to press the button on the laser cutter to start shaping the first piece of the steel hull. That's what "cutting steel" means.

It's worth a celebration, so why not make a milestone payment?

Unfortunately, the contractor stuck its hand up for this particular milestone barely two months after signing the contract. Yes, it had cut steel. No, the design wasn't finished, not by a country mile.

Cutting the steel was not a sign that anything useful had been achieved, it just meant the contractor was desperate to pocket the £2.8 million of the milestone payment.

And construction went downhill from there. The first of the two ferries, originally scheduled for delivery in 2018, didn't start in service until January 2025 and the second is still a work in progress.

Milestone payments can be a really useful incentive, helping to smooth cash flow and reduce finance costs for a contractor.

But you do need to be a wee bit careful with the definition.

Small symptom, big problem

I once took over a project on a temporary basis while a permanent replacement was found for a newly-departed project manager. Day one, I did a walk around the floor and knew at once the project was in trouble.

There was a humming electrical cabinet stuffed full of wiring. The door was obviously intended to be closed, but that wasn't possible because a cable from the cabinet led off somewhere else. Irregular but probably harmless, except for one thing. The kitchen area was a long way from the project space, so the team kept filled water jugs on hand. Yes, you guessed it, they kept them on top of the open electrical cabinet.

A little thing, but symptomatic of the bigger problems the project was facing – morale was low and people either hadn't noticed an obvious safety issue or couldn't be bothered doing anything about it.

But the response was encouraging. I didn't even have to say anything. The guys realised what I was looking at and immediately moved the jugs to a nearby table. When I came back the next morning, the obstructing cable had vanished and the electrical cabinet door was properly closed. A good first step to getting the project back on track.

Little things matter.

Collaboration

Why do we need "collaborative contracts" to behave collaboratively?

Signing a contract labelled "alliance" doesn't create an alliance. Successful collaboration always involves investing time and effort in building a relationship.

That time and effort is worthwhile no matter what sort of contract you have.

Let's face it, uncollaborative behaviours (lying, bullying, shooting messengers) tend to deliver moderately vile results no matter what type of contract gets signed up to.

I'm all in favour of collaboration. Just not of restricting it to "collaborative contracts".

Completion is not enough

Delivering the benefits of a megaproject requires more than just finishing the build. Ask the people who commissioned the Millennium Dome.

The Dome did get completed in time to fill its role as an exhibition venue to celebrate the millennium: it opened on 31 December 1999 (okay, traffic chaos made it difficult for those who had scored a ticket to get there, but it was open) and the exhibition remained open for every day of 2000, closing on 31 December.

Alas, the exhibition made a loss, attracting barely half the predicted 12 million visitors, despite the Dome featuring in the James Bond movie *The World Is Not Enough*. An initial competition for finding alternative legacy uses was unsuccessful. Let's face it, a snapshot view at the end of 2001 would have declared the Dome to be an expensive white elephant. And if everyone had given up at that point, it would have been.

Fast forward to 2024 and things look a bit different. In 2002 the government finally managed to sell the site for development as an entertainment venue, eventually named The O2. The venue has become one of the world's most popular destinations, attracting around nine million visitors a year.

According to the National Audit Office, it has brought employment to the area and achieved *"far-reaching benefits and development across the Greenwich Peninsula"*.

Which is what was hoped for from the project in the first place.

But it wasn't enough just to build it.

The nuclear option

In the face of an abysmally poor decision by a Minister, what can you do?

Transport Scotland faced just such a decision in 2007, when the new SNP government was considering whether to proceed with the Edinburgh Tram project initiated by the previous government.

The project was being led by **tie**, a subsidiary of the Council of the City of Edinburgh (CEC). Governance was handled by a Trams Project Board. Transport Scotland's Director Rail Delivery was a member of this board, and Transport Scotland had lent much of its expertise to the development of the project.

The government's contribution was to be £500m, CEC had agreed to contribute £45 million, and **tie** had assessed the cost of the Edinburgh Airport to Newhaven line at £501.8 million, meaning the funding appeared adequate. The subsequent Inquiry noted the new government agreed to go ahead, provided its contribution was limited to the promised £500 million and CEC took on liability for any cost overrun.

And that was where it started to go pear-shaped. Apparently terrified that liability might fall back on government, the Cabinet Secretary wanted Transport Scotland to "*scale back*" its involvement and gave "*very strong signals*" that it should cease to be represented on the Trams Project Board.

The Director Rail Delivery was appalled. **tie** had relied extensively on Transport Scotland's expertise in developing the project thus far and this expertise and continuity would be lost to the project. He was so horrified he suggested seeking a formal direction from the Cabinet Secretary.

For those of you unfamiliar with public sector practice, this may not sound like much, but the director quite rightly referred to it as the "*nuclear option*". Transport Scotland would effectively be saying, "Minister, what you have proposed is so unutterably stupid we won't do it unless you give us a legally binding direction in writing, which will let us off the hook and put you in the dock when the disaster eventuates."

They wimped out. Didn't do it.

Okay, it would probably have been a career-limiting move.

On the other hand, it might have prevented the fiasco the Edinburgh Tram project became – three years late, two kilometres short and hundreds of millions over budget.

It seems nobody thought to say, "That's a very *courageous* decision, Minister." Where was Sir Humphrey when they needed him?

It's the election, stupid

Not every election determines the future of the free world. The desire of governments to get re-elected does, however, have all-too-frequent negative consequences in the world of megaprojects.

Back in 2014, the NSW government decided it absolutely had to award the $2.1 billion PPP contract for the Sydney light rail before the 2015 state election.

The decision meant there was no time for Transport for NSW to do a number of things that would have significantly improved the prospects of bringing the project in on time, on budget and without disputes. (Score: 0 out of 3, final cost somewhere north of $3.1 billion.)

The biggest problem, as often with light rail, was the utilities buried along the route. TfNSW carried out only a limited discovery of utilities. While they did find more than 2,000, the contractor later turned up another 1,759, so it wasn't exactly a comprehensive survey.

TfNSW also did not reach any agreement with the utilities providers on the treatment that would be required for the utilities discovered, nor on a process for reaching agreement. The PPP's design and construct subcontractor later sued, alleging misleading or deceptive conduct around the provision of information on how to manage electricity cables on George Street. (The lawsuit was eventually settled.)

TfNSW was aware utilities relocation posed a significant construction risk, but like many a government agency in a similar position, judged it was more important to get the PPP contract signed before the election than to take further steps to mitigate the risk.

What they perhaps had not bargained for? In addition to the costs and delays of the utilities relocation itself, they were also hit with a successful lawsuit from businesses along the route whose revenues had been cruelled by the length of time the roadworks remained outside their doors. The NSW Supreme Court ruled that the extended period of disruption constituted an unlawful nuisance, going beyond what was reasonable.

The election result? Yeah, they won.

Sued by the winner

When putting together a procurement process, the tendency is to focus on not getting sued by the losing bidder. But you're much more likely to end up in court with the winner.

The City of Sacramento is probably going to find that out fairly soon.

They developed a new walking and cycling path, the Del Rio Trail. The route crossed a major highway, Interstate 5, so Sacramento commissioned a new overbridge to allow safe travel for pedestrians and cyclists.

The trail opened in May 2024. The bridge turned out to be unsafe, and was closed within months. The problem appears to be the use of unsuitably light concrete and rebar. It isn't clear from the press reports where the fault lies, but litigation seems a likely outcome.

Losing bidders can only sue in relation to the procurement process itself.

Winning bidders are still there when the procurement process is over and they sign up to a tableful of contractual documentation which opens up a whole new range of opportunities for the parties to sue one another.

You can run a procurement process which is impeccable from a probity point of view. But if you don't produce a good specification, or appoint an incompetent contractor, or fail to manage the contract properly, the result can still be seriously substandard.

By all means worry about the losing bidder.

But it's the winning bidder you're more likely to end up in court with.

Poor contracts get worse

The thing about issuing a poorly-drafted contract with your request for tenders is that it will inevitably get worse before you sign it.

In case you haven't encountered it before, the standard excuse a lawyer gives for an incomprehensible clause in a contract is: "It was very heavily negotiated."

The more a contract is subject to negotiation, the more incoherent it becomes. Inconsistencies slip in. Links from one clause to another are broken. Clauses added "for the avoidance of doubt" create more doubt than they avoid.

Plus, the more a contract is subject to negotiation, the more expensive it is to produce. Commercial negotiation is labour-intensive, and the labour in question is mostly the labour of people who charge a lot for their time.

The obvious (but sadly rare) solution is to issue a well-drafted contract with a fair allocation of risk in the first place, so the bidders don't have anything to object to. You probably won't get away with no negotiation at all, because megaprojects are just too complicated. But you can get close.

So why do governments put forward contracts with unreasonable risk allocations? Bespoke contracts instead of industry standard contracts? Contracts with inconsistent or ambiguous specifications? Contracts with key sections left open for discussion?

Setting yourself up for prolonged negotiations just puts up your costs and delivers a poorer outcome.

The world moves on

Doing what worked before is a great solution until it isn't.

When RailCorp, the then operator of Sydney's railway, signed a contract in 2006 to buy 78 trains' worth of rolling stock, the contract also included a requirement for simulators to enable crew training.

RailCorp had for some years used simulators to help train the crew, and it was clearly going to be necessary this time round because there were so many new trains that hundreds of drivers and guards would have to be trained in a relatively short period of time.

Everyone was happy with the existing set-up, so the old simulator specification was dusted off and pressed into service for the tender process.

It didn't at the time occur to anyone that the technology had moved on a bit. The existing simulators for the old Tangara fleet used some fake routes to introduce the drivers to standard situations they might encounter, and that was included in the specification.

It was only later that someone from the supplier mentioned they could (for a fee) produce simulations of actual routes on the network – did we really want fakes when these days we could have real ones?

Ah. Yes. Good idea. Shame we didn't think of it earlier.

The soccer ball

According to *Fluke*, a recent book by Brian Klass, life is a great deal more random than we're conditioned to think it is. (I'm not sure this is total news to project managers, most of whom would tell you that dealing with random crap is more or less the day job.)

One of the stories Klaas tells is of a tourist named Ivan, who in 2022 was caught in powerful currents off Myti Beach in northern Greece and swept out to sea. He was saved because he happened on a floating soccer ball, later claimed to have been accidentally kicked into the sea a couple of days previously by a small boy eighty miles away. Ivan clung to the soccer ball and it kept him afloat long enough to be rescued.

Which is about as random as it gets.

Klaas looked at the incident and marvelled at how your smallest acts can have life or death impacts you may never hear about.

Yeah. My reaction was more, "How did it happen in the first place?". Was he swimming between the flags? Were there warning signs up about the currents that he ignored? Did he not know anything about swimming safety?

Okay, maybe some butterfly is going to flap its wings at the wrong moment and you'll die anyway, but you can seriously increase your chance of survival by following safety procedures and wearing your PPE.

Or you can trust to luck and hope for a soccer ball.

The menace of silo budgets

Railtrack, predecessor to Network Rail, used to have a system where project feasibility studies came out of the operational budget. Unless the project went ahead, in which case the cost came out of the capital budget instead.

Which was an absolutely splendid incentive for operations managers to push ahead with projects of marginal or insufficiently-developed feasibility, just to get their budgets back.

Equally, the capital project managers didn't worry if their corner-cutting pushed up the maintenance costs, because that came out of someone else's budget.

The budgeting system wasn't the primary cause of Railtrack having to be put into administration, but it certainly didn't help.

Silo budgets. Bad news.

Client decision or bidder decision?

Should contract trade-offs be made by the client or the contractor? The standard answer, that the party taking the risk should be making the trade-offs, doesn't always work.

We ran into this issue on the Waratah train procurement. The supplier was going to be responsible for maintaining the trains for 25 or so years. They only got their full availability payment for a train if it was functioning properly, so the quantum of spares they held at the maintenance facility during the operating phase was clearly their risk. Was it their trade-off?

More or less. The problem child was not the spare parts but the spare trains.

A lot of maintenance tasks can be slotted into gaps in the timetable, but there will always be some that require a train to be taken out of service. To ensure the supplier could meet its contractual obligation to have 72 trains available for service each day, there would need to be a number of additional trains to cover the maintenance requirement.

Who should pick the number?

In theory, the contractor would be taking the risk, so the contractor should make the decision. Not that simple. Pretender calculations by the client indicated the optimum number of spare trains was five and a bit.

Only problem, you can't have a bit of a train. And a whole train is expensive.

Bidders would have to choose between offering only five spare trains, meaning their bid would be cheaper but they would struggle in the operational phase, or offering six, which would keep them well covered during the operational phase but would put up the cost of the bid so they might not win the contract at all.

And that was not a trade-off we felt it was safe to let the bidders make.

Solution? We took the decision out of their hands by specifying a minimum of six spare trains.

Which project management system?

I never had any formal training in project management.

To this day I couldn't tell you the difference between Agile and Waterfall and I would probably believe you if you told me PMBOK was a type of South African deer. The first time I received an email headed "SCRUM TRAINING", I deleted it unread and wondered which joker in the team had put me on a mailing list for rugby players.

I do know that all the different project management systems have enthusiastic proponents, and each one can point to successful projects. So how do you know what system is best?

There was a productivity experiment done many years ago. I forget the precise details, but it involved re-painting the white walls of an office in a colour (green?) that was supposed to promote productivity. Productivity duly went up. Then they found an office which was already painted green and re-painted it white. Productivity also went up. People just work better when they see the company values them enough to re-paint the office. Colour not relevant.

So I do wonder, while freely acknowledging my ignorance, if the different project management systems are a bit like that. There's a lot of basic stuff you need to do to keep a project on track. Pinning down the scope, getting decisions made when you need them, keeping tabs on the budget and schedule, keeping every-body in the loop. Etc. And the different systems all have processes for making sure this stuff gets done.

When a company adopts a new project management system, it comes with the blessing of the CEO, a star-spangled roll-out by champions and training for everybody. So, when things improve is it really the new system, or is it just that people do better when ... they have the blessing of the CEO, a star-spangled roll-out by champions and training for everybody?

Mothballs

Sydney's desalination plant spent almost 7 years out of use.

When the NSW government in 2007 committed to build it, the dam storage was 34%, scarily low and well under the official trigger level of 48%.

The plant was finished in 2010 but the dam levels were then high and after a two-year proving period the plant was mothballed, finally returning to production in January 2019.

It takes time to set up a megaproject. Although at the decision point the dam levels were at 34%, by the time the contract for the desalination plant was actually ready for signature, storage had already gone back up to 57%, well above the trigger level.

In a later report into urban water supply, the Productivity Commission pointed out that "Large savings are likely to have been available to the community if the government had taken and exercised an option to delay construction, even if this option incurred costs."

They were suggesting, in other words, that it would have been better value to accept the expense of mothballing the contract now and build the desalination plant later, rather than to go ahead with the contract now and incur the expense of moth-balling the plant later.

It would have been a difficult call to make. The momentum

behind a megaproject contract at point of signature is almost unstoppable.

But if the basic rationale for a project has changed, whether to mothball the contract is at the very least a question that should be asked.

PART 10

More haste, less speed

An impossible schedule is a recurring theme in megaproject disasters. But the Scottish ferry procurement is a classic.

The decision to buy two ferries to improve the connection of the Scottish Isles to the mainland, at an estimated cost of £40 million each, was made in July 2014. At the time, it was anticipated the contract would be awarded at the end of March 2015.

Given there was no specification in existence at the time of the decision, it's hardly surprising this schedule could not be sustained.

But for a while there, they tried.

Caledonian Maritime Assets Limited (CMAL), the Scottish agency which owns the Scottish ferries and ports infrastructure, was charged with procuring the new ferries. A senior executive of Caledonian MacBrayne (CalMac), the ferry operator, later said that for such a procurement they would normally expect to spend up to a year with a design consultant, working up a detailed specification.

CMAL gave them three weeks.

And yes, of course, the procurement schedule was extended because it was just plain impossible. Duh. But too late to stop the request for tenders going out with the cobbled-together-in-three-weeks specification.

CalMac complained that the specification only contained about 20% of their requirements. It didn't get sorted out before contract award in October 2015, which inevitably led to schedule-killing variations. (The contractor going bust didn't help the schedule either.)

The two ferries were due to be delivered in 2018. The first one made it into service in January 2025. Scotland is still waiting for the second.

How to be commercial

The break-up of British Rail in the 1990s resulted in a sudden demand for contract managers. Almost from one day to the next, people who were used to interacting as employees of the same company were being asked to interact as client and contractor.

They weren't very good at it.

I was working at Railtrack when the transition happened, having been brought in to write the prospectus for the flotation. It was pretty grim.

Collaborative contracting was not then a thing. Collaborative? Those contracts were about as collaborative as a bullet in the head.

People used to a public sector environment suddenly felt under enormous pressure to be "commercial". Unfortunately, the only technique most of them knew was to enforce the terms of the new rail access contracts absolutely strictly, regardless of whether this achieved any commercially useful outcome.

The most visible result of this approach was the creation of a small army of rail employees to administer the onerous performance regimes: their job was not to reduce delay but to argue about who was responsible for it.

"Commercial"? Not very.

To me, being commercial means delivering good commercial outcomes. Which may or may not involve strict enforcement of the contract.

Rocky recommendations

If you want to avoid a legal challenge on an evaluation, it's essential the recommendations to the decision-maker are supported by the material presented in the report.

Every now and again, someone will forget to put their brain into gear and will decide to fix a mismatch by omitting relevant material from the report instead of by changing the recommendation. True, the revised report will appear to support the original recommendation, but only because the report is now misleading. That won't protect the decision against legal challenge.

Even if nobody sues, you can still get some pretty sub-standard outcomes. When the Australian Civil Aviation Authority tried to procure an air traffic control system in the 1990s, the board had to decide how many bidders to shortlist.

The evaluation report contained a recommendation to retain three rather than two bidders for the final stage. There was a table showing the relative ratings of the bidders in specific areas, which demonstrated the third bidder had no chance of winning the contract. This highly relevant information did not support the recommendation to proceed with three bidders rather than two. It was omitted from the report.

The recommendation was accepted, further submissions were requested from the three bidders, and ten wasted weeks later the third bidder was duly eliminated.

Why did they do it? Well, the agency had decided to request pricing in the initial round, and the third bidder's price was exquisitely tempting. In the absence of defined commercial terms, the prices were of course totally meaningless, and so it proved.

If a properly prepared report doesn't support your recommendation, change the recommendation, not the report.

Whose risk?

It's easy to make a bad decision when someone else takes the risk. Pilots for the US Mail service found a way to turn that on its head.

In the early days of commercial flying after WWI, the risks were appallingly high. Paul Carroll, in his book *Billion Dollar Lessons*, writes that "*Of the first forty pilots with the US Mail service, thirty-one died carrying the mail. The life expectancy of a pilot was four years.*"

One of the issues was that planes in those days couldn't cope with much in the way of bad weather. And the managers of the airfields, keen to get the mail delivered, weren't very sympathetic to pilots who thought it was too dangerous to fly.

In 1922, the pilots managed to negotiate a deal:

"If the manager of a field told a pilot he had to take off to deliver the mail even though the pilot thought the weather was too dangerous, the manager had to be willing to sit in the plane's second seat and fly once around the field."

The enforced risk sharing did wonders for management's attitude. There were no fatalities that year.

Lessons delayed

There was a lot wrong with the Edinburgh Tram project, but it is noteworthy that the first four recommendations in the Inquiry report had absolutely nothing to do with the project.

How could that happen? Commissions of Inquiry have a closely defined remit, and are not entitled to step beyond it. What was going on?

The clue is in the opening words of the first recommendation:

"Scottish Ministers should undertake a review of public inquiries to determine the most cost-effective method of avoiding delay in the establishment of an inquiry..."

The contract for the Edinburgh tram works was let in May 2008. The line was supposed to be finished in 2011. After extended delays and cost overruns, a truncated version of the line (about two kilometres short of the originally intended destination) was eventually opened in May 2014. One month later, in June 2014, the Commission of Inquiry was announced with Lord Hardie as Chair.

More than eight years and £13 million later, Lord Hardie's report was finally delivered on 19 September 2023.

As a lessons-learned exercise, the report is exemplary, but it wouldn't have hurt anybody to have those lessons learned a bit earlier.

Improving the establishment and delivery of public inquiries may have been outside his remit, but Lord Hardie was spot on with the need to do something about it.

Stupid outcomes not required

Whatever a probity auditor may tell you, if something would deliver a stupid outcome, probity does not require you to do it.

Ordinarily, for instance, it is totally unacceptable to change the evaluation criteria or methodology partway through a procurement process. Particularly if bids have already been submitted. But rules can't cater for every possible thing that can go wrong.

On one contract I dealt with, we ran into a major issue while evaluating the bids. Both the shortlisted bidders had produced a price that was way above the anticipated range – we're talking $50 million plus.

Problem. Treasury was never going to cough up the extra $50 million.

Second problem. We had absolutely no idea why the bidders seemed to be playing in a different ballpark.

So we called a temporary halt to the procurement and told the bidders what our problem was. It quickly became apparent we had included a requirement for some functionality we had thought was standard. It wasn't. What was more, it would require some pretty complicated software development to add it to the package. Hence the extra $50 million.

There was clearly no point in continuing the process as it was. It was suggested probity required us to bin the procurement and start again at the beginning. But there was a lot of sunk cost for both us and the bidders, and it seemed stupid to have to go back and incur it all again. Did we really have to?

No, of course not.

After further consultation (including with Treasury), the operator agreed it could manage without the non-standard requirement. The specification and the evaluation methodology were re-jigged accordingly, and the bidders were given some extra time to submit fresh bids. They were totally okay with the process change (and yes, we did get that in writing) and the probity auditor was able to give us a clean report.

Easy.

Passive aggression

The London Garden Bridge project, initiated during Boris Johnston's flamboyant stint as Mayor, swallowed about £43 million of public money while failing to build anything at all. The award of the two significant contracts on the project was, shall we say, not entirely in accordance with best procurement practice.

Dame Margaret Hodge MP ran an independent inquiry into the project in 2017. One paragraph of her inquiry report caught my eye, in the section discussing successive drafts of an earlier report into the contract procurements by the internal audit team:

"Elsewhere the criticisms made by Internal Audit were watered down. For example ... when discussing how [the winning bidder] came to be interviewed when they had not scored well, a sentence that originally said: "TfL Planning requested that [the bidder] should be interviewed" was amended to: *"a decision was taken to interview [the bidder]."*

Gotcha!

Proponents of plain English complain about the use of the passive voice, saying it makes sentences easier to understand if you use only the active voice. This totally misses the point. Public sector use of the passive voice is not caused by failure to comprehend the mechanics of good communication.

The issue is that if you use the active voice instead of the passive to say, "it was decided", you have a problem. The equivalent phrase, "so-and-so decided" requires you to a) know and b) disclose the identity of the so-and-so who made the decision.

Perish the thought.

Public sector employees know perfectly well that failure is an orphan and success has many fathers. They just don't know in advance which one is going to apply. Routine use of the passive voice enables acceptance of responsibility for a decision to be deferred until the advisability of acceptance has become apparent.

And when, as here, it is already apparent that the fan is about to distribute noxious substances, what do they do?

"A decision is taken" to use the passive voice.

Another classic demonstration that the public sector has elevated the avoidance of accountability into an art form.

A capital solution

It's always fun to spend money on capital projects, but sometimes operational solutions are a better option.

There is a story in the rail industry, possibly apocryphal, of two railway companies that both had a major problem with vandalism. Rock-throwing meant broken windscreens, and replacing a broken windscreen was a headache.

The replacement could only be done at a maintenance facility, and while the actual replacement took less than two hours, the train could not be moved until the adhesive was fully set, which took another 30 hours. The vandalised trains clogged the maintenance facilities and caused serious delays in planned fleet maintenance.

The first train company solved the problem by building, at a cost of some millions of dollars, a dedicated covered road where vandalised trains could be repaired without interfering with regular maintenance. The new road was completed on time, on budget, and the rail company was very pleased with the outcome.

The other rail company went to their supplier of windscreen adhesive and asked, "Could you produce a quick-drying adhesive?"

Yes, they could. Problem solved.

The joys of litigation

What's the difference between being sued and being the subject of a public inquiry?

They have a lot in common, none of it good.

You spend hours and days of time you don't have scouring your memory and the files trying to work out what happened.

You realise that people who seemed to be totally supportive at the time now have a completely different recollection of events.

And you get subjected in public to thorough and unnerving cross-examination.

Unpleasant as both experiences are, being sued is probably better. There's always the possibility of being able to negotiate an acceptable settlement, and if it does go to court, you might win.

But a public inquiry? Government doesn't initiate one unless you've already lost.

Megaprojects change the landscape

Occasionally you get a megaproject where the new infrastructure is completely invisible. A sewage main, for instance, will probably not come with a visitor centre.

But when you are talking megaprojects, invisibility is rare. The scale of construction means that, for good or ill, the new infrastructure becomes a landmark. Which leaves project planners with decisions about where and how to make the architectural statement.

In the UK, the rail link from St Pancras to the Channel Tunnel has two: St Pancras Station and the bridge across the Medway. Both impressive.

In Sydney, the new Sydney Metro made big architectural statements at each of the city stations. Central is my favourite, but they're all pretty amazing. The opening in 2024 gave a big lift to the city.

But I still haven't forgiven Sydney for the Epping-Chatswood Rail Link, now absorbed into the Metro. The original design involved putting part of the route between Chatswood and North Ryde on a viaduct through the Lane Cove National Park. It could have been spectacular.

Instead, we got a ridiculously long and expensive tunnel that had to be dug around a huge curve away from the direct route to achieve enough depth to get under the Lane Cove River. Permanently longer journey times, more wear and tear on the trains, and ongoing maintenance to keep the Lane Cove River out of the tunnel.

A missed opportunity.

Lacking assurance

Sometimes committees just don't get it.

The classic was the Contract Award Committee on the InterCity West Coast Franchise debacle back in 2012, keenly analysed in the Laidlaw Inquiry.

The Contract Award Committee's terms of reference said its role was *"providing assurance on the procurement process"*. (It was a £5.5 billion contract so having somebody provide assurance was clearly a good idea.)

The Committee didn't meet at all for five months following the launch of the invitation to tender, thereby failing to notice that there was no Senior Responsible Owner on the project, the sort of omission an assurance process is supposed to pick up.

When it finally did meet, it unilaterally decided its role was to determine the amount of subordinated loan facilities to be required of the bidders. It determined these amounts by:

a) feeding some parameters into an unaudited financial model which had been designed for a completely different purpose;

b) failing to notice that the model outputs were in real rather than nominal currency; and then

c) deciding it didn't like the outputs and making up some new numbers using a discretion the bidders had been told it didn't have.

Oh, and the numbers it made up increased the loan facility requirement on one bidder from £0 to £40 million and reduced the requirement on the other from £252 million to £190 million. No biggie.

At the meeting which approved the award of the contract to the preferred bidder, subject to confirmation by those further up the inadequately-documented approvals chain, it gallantly discharged the assurance function in its remit with an exchange between a committee member and the project team leader.

The committee member "*sought assurances that the ICWC procurement had followed the DfT's published procurement process and would be robust in the face of any challenges*".

The assurances were provided by the project team leader, a luckless individual who had been put in charge of procuring a £5.5 billion contract, despite being only in Pay Band 7 (something achieved by a friend of mine at the ripe old age of 24), and therefore definitely not senior staff.

He said everything was fine.

To be strictly accurate, the minutes record he confirmed that "*the process remained robust and that any issues raised would be more focussed around policy than process*".

Shortly afterwards, the losing bidder brought an action for judicial review and the whole thing had to be cancelled, at a cost of about £40 million in bidder compensation.

The Olympic swimming pool

How did civil engineering projects manage before Olympic swimming pools?

The Paris Olympics of 1924 was the first to introduce the 50-metre swimming pool with marked lanes. That size pool is now not only the Olympic standard for swimming races, it is also a standard measure of volume. You see it everywhere:

- The Sydney Metro West project in 2022, reporting on a new Metro station planned at Sydney Olympic Park, announced that "*To construct the station box, a total of 468,000 tonnes of rock and soil will be excavated, equivalent to 78 Olympic swimming pools*".
- In 2023 we saw the headline: "*NASA says an asteroid the size of an Olympic swimming pool could hit Earth in 23 years*".
- The *BullionByPost* website today says that all the gold ever mined is "*just less than enough to fill 3.5 Olympic swimming pools*".

But when the Panama Canal was built, in the early years of the 20th century, the Olympic swimming pool was not yet a unit of measurement. How was an English engineer writing in 1915 to describe the 200,000,000 cubic yards of earth that had to be shifted to build the Panama Canal?

One Gordon Knox gave it a go: *"The removal of this quantity of material from the Isle of Wight would lower the entire surface of the island by 15 inches."*

I guess that makes it official. If you don't have an Olympic swimming pool, use the Isle of Wight.

Whose turn to be god?

Government is god. It's in the contract. Although sometimes the wording is a bit different.

In our secular age, lawyers writing the force majeure clause can now find other ways of describing what used always to be known as "acts of god": those natural disasters where it is considered unreasonable for the normal risk allocation to apply.

However they are described, in a public sector project the contract almost always provides for the consequences of those acts of god to be laid at government's door, as if the flood or war or hurricane or whatever was an instruction from government rather than divine intervention.

So if god does it, the contractual effect is the same as if government did it. Even if government isn't god, it might as well be.

It's a clause which probably needs a bit of work before you use it. These boilerplate clauses come straight out of the lawyers' precedent bank, and the problem is that what would be an act of god in a small procurement may well be business as usual for a megaproject.

I haven't seen the Thames Barrier contract, for instance, but you can be sure it didn't let the contractor off the hook for high tides and twenty-year storm surges, because the whole purpose of the Thames Barrier is to protect London against such events.

Electrical storms in Sydney are a regular feature of summer weather: a contractor building communications towers needs to build in protection against lightning strikes, not to be held harmless when a strike occurs.

On a megaproject, sometimes it's the contractor's turn to be god.

Your call

Any megaproject director will from time to time find themselves asking, possibly in tones of extreme exasperation, "So whose idea was *that*?"

In theory, all the key decisions around a megaproject should be taken at the appropriate level and recorded in a decision register, or at least in the minutes of the relevant committee.

Yeah, right.

I'm assured by someone who ought to know that the Sydney CBD Metro shambles back in 2008, when the state spent about half a billion dollars failing to build a metro, happened because the idea was mentioned in a memo to the premier. Instead of asking for further information, the premier unilaterally announced it at a press conference, to the total bemusement of the attendant bureaucrats. Apparently nobody cared to contradict him, so the project was duly launched and continued until it was crushed by the weight of its own stupidity.

Public sector decisions can appear without trace. Perhaps someone forms a view that the Minister wants whatever it is.

Perhaps an idea emerges from a discussion among the ignorant and spreads by osmosis. Who knows?

The issue for you as project director is that if the idea has made it into the project, then if you don't have a record of the decision being made by the steering committee, or somebody else with appropriate authority, that decision is yours. You're in charge. Nobody has told you what to do. (Or not so you can prove it!)

If you can't demonstrate whose idea it was, then guess what. It was yours.

Too many holes, not enough cheese

Looking again at the Clapham Junction rail accident report, I realise this was a classic case of Swiss cheese defences: layers of protection, each with holes in it. The holes lined up on 12 December 1988 and 35 people died.

The accident occurred because a signal that should have been set at red instead showed green. A train went through the signal. It ran into the back of a stationary train and derailed into the path of a third train approaching from the other direction on the adjacent track.

The signal had been rewired the day before. Without going into the electrics, the rewiring required a single wire to be replaced with two new wires. The old wire was pushed out of the way. It should not have been able to reach the old contact point, but it moved back to its original position and made contact, keeping the signal green.

There were four separate defences against contact being re-made:

1. Tying the wire back, so it could not move from the out-of-the-way position.
2. Cutting the wire short, so it could no longer reach the contact point even if it did move.
3. Wrapping the tip of the wire with new insulating tape, so even if the wire did move and did make physical contact it would not create an electrical connection.
4. Detaching the other end of the wire from the fuse, so no electrical contact could be created even if the first three defences failed.

Good enough, you would have thought. Any one of these defences would have been sufficient on its own. And to back up the last one, an independent "wire count" should have been done to confirm the other end of the wire had been detached.

The signal worker was thought to be efficient and competent, had been doing this sort of job for sixteen years. No one had ever told him an old wire should be tied back, so he hadn't done it. Ever. That piece of cheese had a very large hole in it.

No one had ever told him the old wire should have been cut short, so he never did that, either. Another hole in the cheese.

He did normally use insulating tape – old rather than new, but that would probably have done the job. Only that day he didn't (possibly because he was interrupted during the work, possibly because he had been working seven-day weeks for an extended period) and thereby created another hole in the cheese.

He also failed to detach the other end of the old wire from the fuse. This hole in the cheese should have been patched by an independent check, but that didn't happen either. His immediate supervisor literally failed to get the memo and didn't know an independent check had been required for such works since

May of the previous year. The testing engineer, on a temporary placement, was theoretically responsible for ensuring the independent check was carried out but was also unaware of the requirement.

When the holes to cheese ratio is catastrophically high, catastrophe is what you get.

Benefits of remote working

During the pandemic, I did a review of a regional project in NSW – my first attempt at doing such a review over the internet instead of on the spot.

Surprisingly, the Project Director said that in many ways Covid had made his job easier, because it gave him better access to senior people at HQ.

Keeping his problems front-of-mind with the executive had used to be a real issue. Sydney-based project managers could just walk into the boss's office and get attention. For him, several hundred kilometres away, it wasn't so easy. He would have to fly to Sydney, and probably stay overnight, just to get an hour with his boss face to face.

Then Covid happened. The only way anybody could meet with the boss was during a scheduled video call.

Finally, project managers in the bush were on an equal footing with those based in Sydney. A huge improvement.

Has that been sustained now everyone is back in the office? I wonder.

PART 11

The secret of professional integrity

The key to professional integrity is to spend less than you earn and invest the difference.

??????

Yep.

Okay, "spend less than you earn and invest the difference" is normally regarded as the key to wealth, not professional integrity.

But professional integrity is not just about having high professional standards, it's about being willing to put yourself on the line to uphold them. It is really hard to refuse a dodgy client if the fee is paying your mortgage this month.

It's no accident that criminals seeking to corrupt public officials target the ones drowning in debt or with serious gambling addictions. Their professional integrity is compromised. They're vulnerable.

Integrity requires financial resilience.

So if you want to have professional integrity, spend less than you earn and invest the difference.

Also, you get rich.

Win-win.

The binary decision-making tool

In July 2013 the NSW government embarked on the sale of Macquarie Generation, a state-owned electricity generator, with the announced objective *"to unlock funds for critically needed infrastructure across NSW"*. (Translation: "raise money".)

For the sale to go ahead, the government, in an internal decision, made it mandatory that the sale proceeds should exceed the retention value of the asset. That amount was not made public, so when the asset went out to tender, it was a toss-up whether any of the bids would exceed the retention value.

The thing about tossing a coin, the ultimate binary decision-making tool, is not that heads or tails gives you the right answer, it's the reaction you experience when you do it. You may have thought you didn't care either way, but when the coin is spinning, suddenly you know which way you're hoping it will land.

For the NSW government, it landed tails. Of the three bids received, only one exceeded the retention value, and that one required clearance from the Australian Competition and Consumer Commission. Clearance was refused.

How did they feel when the coin landed? Would they have been willing to accept a lesser bid after all? Should that criterion really have been mandatory?

Whatever they felt, they got lucky. The top bidder appealed to the Australian Competition Tribunal, which allowed the deal to go ahead. The sale was eventually completed in September 2014.

I tend to dislike mandatory criteria anyway, because megaprojects are so complicated it is rare for any one factor to be a showstopper if everything else stacks up.

But if you're thinking of having one? Before you write it into the request for tenders, toss a coin. See how you really feel.

The perfect KPI

Sometimes you don't need good performance, you need perfect performance.

We all had our noses rubbed in that in November 2023, when Optus went into meltdown. Optus customers couldn't make a phone call, couldn't access the internet, couldn't make or take payments, some couldn't even call emergency services. For some Optus customers this toxic package was sufficiently disastrous that they switched providers rather than risk it ever happening again.

Something that triggers termination of a contract is a useless KPI. A good KPI lets you know when performance is slipping, so you can do something about it before performance reaches an unacceptable level. But if only perfection is acceptable, what can you do?

The best I've come across is to find "precursor events", and measure those instead. If you have 24/7 server back-up, for instance, a precursor event might be a server failure that leads to the back-up being activated. The event is invisible to the user, so there is no actual service downtime, but if a second failure were to occur, there would now be no back-up, so the risk of a total outage has just gone up.

It may seem perverse or just plain weird to treat something as a service failure for KPI purposes when there is no failure of service to users. But if you need perfection, you don't have much choice.

Breach of confidence

If you're any kind of leader, someday someone junior will stick their head around your office door and ask if you will keep something confidential if they tell you about it.

Never say yes.

It's tempting, of course, because you're not the sort of person who breaks confidences anyway, so why not agree?

Well, why do you think this person wants to tell you something in confidence? There's a distinct chance they've spotted some problem they don't feel able to deal with and they want to dump it in your lap.

That much is okay. However much we wish our juniors could solve problems on their own, sometimes they need help and as their boss it's our job to see they get it.

But if it was just an ordinary work problem, they wouldn't be asking you to keep it confidential. So you already know there's a can of worms headed your way. Whether the problem is that somebody has just received a cancer diagnosis or somebody else is defrauding the company or whatever it is, once you know about it, your responsibility to the organisation means you will have to take action.

Have to.

The problem? Taking action and keeping confidentiality are sometimes mutually exclusive.

So, the right answer is something along the lines of: "I'll keep it confidential if I can, but I have responsibilities to other people as well and I may not be able to. Do you still want to tell me?" (If the organisation is sufficiently enlightened to have a proper whistleblowing policy, you can also refer them to that.)

The junior will nearly always decide to tell you anyway. They're coming to you because they trust you. Being open about not promising confidentiality will probably make them trust you more.

But don't promise to keep the matter confidential, because it's a promise you may not be able to keep.

Business as usual

What did the collapse of the Francis Scott Key Bridge in Baltimore in 2024 have in common with the Columbia space shuttle disaster of 2003?

The investigation into the collapse is ongoing, so this is pure speculation. But I suggest what they had in common was that a life-threatening situation had become business as usual.

In the case of the Columbia space shuttle, foam insulation coating the shuttle broke off during the 2003 launch, compromising the integrity of the insulation layer: the shuttle burned up on re-entry, killing the seven astronauts on board.

Pieces of foam had broken off on previous flights without serious consequences, but on this occasion some of the debris had hit the shuttle itself, ripping off even more of the foam, and the shuttle did not have enough insulation left to survive the descent.

As Ozan Varol describes in his book, *Think Like a Rocket Scientist*, NASA knew perfectly well that pieces of the foam were coming off in flight. But since there had already been 112 flights in which nothing serious had happened as a result, they didn't see the need to do anything to fix it. So bits of foam come off. So what? Business as usual.

Until the Baltimore bridge was hit by the 300-metre container ship *Dali*, it didn't have a 112-flight safe-operation history, it had a 47-*year* safe-operation history. Ships even bigger than the Dali had been going past the bridge for more than a decade. So what? Business as usual.

There comes a point where business as usual amounts to dancing with death.

Response times

It's common for a megaproject contract to require the client to respond to a design submission within ten business days.

That's no reason to set up a process which sees the response to the contractor go out on day ten. Targeting day ten means if you ever miss the target (and everybody misses targets sometimes) you'll be in breach of contract.

Plus, count the number of design submissions. Can you save a week on each one? Even if you only manage it fifty percent of the time, that's a lot of weeks. Anything that relieves pressure on the schedule is a big plus.

Yes, you could regard that time as a gift to the contractor and no, you're not in the business of making gifts. But this is about what's best for the project. Besides, if you're taking ten days to do what you could do in five, that's your own time you're wasting.

If you want collaborative behaviour from the contractor, offer collaborative behaviour in return.

There is very little a contractor values more than quick responses from the client.

Who made the decision?

All too often, nobody knows. Probably it just emerged from some primordial swamp.

You see it time and again in public inquiries. The critical decision that sent everything careering off in the wrong direction is an orphan.

Okay, people probably aren't going to be queuing up to take responsibility for a decision which in hindsight was irretrievably stupid, but in these days of email archives the absence of evidence is striking in itself.

The misguided decision to adopt a prime contractor model for the Queensland Health Payroll System – no evidence. The bizarre decision in the InterCity West Coast franchising decision to calculate guarantee requirements using a financial model that was manifestly not fit for purpose – no evidence. Poor decisions around hotel quarantine in Victoria during the pandemic – no evidence.

This is sadly typical. It stems from a defensive culture: everyone's backside is protected because nobody can identify which backside to kick. Unfortunately, this method of rear end protection is incompatible with successful project delivery.

If you can't identify decision-makers, you can't identify who is responsible for doing things. If you can't identify who is responsible for doing things, things don't get done. And if things don't get done, projects fail.

An ingenious financial model

A supplier once provided me with an impressive set of spreadsheets supposedly demonstrating it could carry out hundreds of millions of dollars' worth of maintenance activity 25 per cent more cheaply than the agency was currently achieving. They knew we were contemplating putting the work out to tender and hoped instead to persuade us to pursue a negotiated outsourcing deal.

The beautifully presented spreadsheets essentially consisted of line items of maintenance expenditure with, for each, an estimate of the current amount being spent on that item, and an estimate of how much more cheaply the contractor could deliver it.

I was interested to know where they thought savings could be made. On closer examination, the spreadsheets left me none the wiser. There had been absolutely no effort to consider where maintenance savings might genuinely be achievable.

Instead, they gave an estimate of spend for the current activity, then applied a 75 per cent multiplier throughout. Not so much ingenious as ingenuous. Or perhaps more likely disingenuous.

We decided to pass up the opportunity.

Not legally binding

An agreement doesn't have to be legally binding to be effective.

The final stage of the Sydney Gateway project, connecting Sydney Airport to the motorway network, opened in 2024: successfully delivered on schedule, despite having been under construction through extreme wet weather, the pandemic lockdowns, supply chain disruption and materials hyperinflation.

One key to its success was the decision to use a Dispute Avoidance Board, a team of three independents who met with the parties quarterly throughout the project with the express purpose of encouraging them to sort out issues early and deal with them before they became intractable problems.

Dispute Avoidance Boards in Australia already have a splendid record in enabling parties to sort out contractual problems without recourse to the horrors of construction litigation or arbitration.

What was special about this project was that in addition to the contractual requirement to engage with the DAB, the parties made a commitment to resolve all issues within three months. Not legally binding. Not enforceable. But a commitment all the same.

And what the commitment meant was that an issue on the list at one quarterly DAB meeting should have been dealt with and off the list by the time of the next one. If it wasn't, the parties knew they would have to suffer the embarrassment of having failed to meet their commitment, not to mention suffering interrogation by the DAB members. It also meant the DAB members knew there was a problem and could help find ways to resolve it.

Not legally binding. Not enforceable. But very, very effective.

Consulting the operator

Infrastructure operators always complain that the people running the infrastructure procurement process fail to consult them properly.

They are quite often right. This is yet another aspect of the procurement process that starred in the Scottish ferry fiasco (two ferries, seven years late and counting, way over budget).

The procurement for the ferries was run for the Scottish government by Caledonian Marine Assets Ltd (CMAL). CalMac Ferries, the operator, was given just three weeks to say what they needed. They later complained the specification that went out with the request for tenders only contained about 20% of their requirements.

When the original list of six bidders had been whittled down to three, CMAL handsomely asked CalMac to review the remaining designs. CalMac's naval architect found the design by a Polish shipyard best matched its requirements, with Ferguson Marine, the only Scottish bidder, ranked second.

This ranking would have ruled the Scottish bidder out of contention, which no doubt explains why CMAL's own assessors then discounted CalMac's assessment, scoring the Polish yard as "poor" and Ferguson Marine as "good".

Scotland triumphant! Shame about the total horlicks they made of the ferry delivery.

Not much use consulting the operator if you then ignore what they say.

A matter of clarification

Bids on major projects are complex, with different experts preparing different parts of the bid.

While it is entirely the responsibility of the bidder to make sure the bid is internally consistent, a few hiccups here and there on a megaproject are probably inevitable. And if the god of random numbers hits the wrong button, a hiccup will go nuclear.

Way back when, I was running a major infrastructure procurement. One bidder included an absolutely outrageous provision in the contract markup, which allowed them to develop software at our expense (whether we wanted it or not) and sell it to other people without paying us anything.

Guess what? That bidder didn't win.

There were other factors, of course, and the evaluation would have passed any challenge. But I think if I were faced with the same situation today, I might approach it differently.

Today, I think I would issue a Request for Clarification. That could be regarded as a bit of a stretch, since the legal provision the bidder had added was utterly clear, so there wasn't anything to clarify. The Request for Clarification process is not there to allow bidders to improve their bids after the deadline.

Nonetheless, the provision did sit rather oddly with the rest of the bid, and looking back I can't help thinking that the bidder's legal team had perhaps gone off on a frolic of their own, which the bid director had simply failed to pick up.

But I will never know, because I didn't ask.

Self-incrimination by email

The existence of email has proved of inestimable value to people wanting to shoot themselves in the foot.

You see it in so many public inquiries into project failures. All those careless emails disinterred from the archives.

The Edinburgh Tram Inquiry found one from a contractor to the client noting he had "*pockled the spreadsheet*" as requested. "*I doubt they will notice what I have done*", he added, thereby inadvertently confirming the funder was being deliberately misled about the cost.

And then there was the Inquiry into the failed Queensland Health payroll system, which found an email from a member of one of the bid teams recording his diligent but unsuccessful attempts to access confidential material mistakenly placed on a public drive.

"*Looks like we were just a little bit too late,*" he wrote, oblivious to the possibility that his actions might be open to criticism.

Not much use denying your involvement if you're the one putting it on the record.

Playback time

Ever suffer from parties taking aggressive stances in a contract negotiation about things the other party never said in the first place?

Me too. The single most effective phrase I have ever used in a negotiation is, "Let me play that back to you." I use it all the time.

Well, actually I get a bit bored with the one phrase, so I like to vary it a bit: "Let me check I heard that correctly." "So what you are saying is ..." "May I just see if I have understood you?" "So your position is ..." "I think I know where you're coming from, but let me check."

I have found this simple tactic has amazing benefits:

- If you know you are going to summarise the other person's words back to them, you are more likely to listen carefully to what they say. This generates goodwill you are bound to need sooner or later.
- If you are listening carefully, you are more likely to understand what they have said, which is critical to achieving the genuine "meeting of minds" that a good contract represents.
- If you have not understood, you will find out immediately, eliminating the fruitless and acrimonious contests generated by reacting badly to something the other party never intended.
- If you have understood, but need to negotiate a different position, the other party will know you are disagreeing with them because your position is genuinely different, rather than assuming it was because you didn't understand what they said. This saves time and makes them much more likely to listen to you.

I've encountered pretty much every negotiating tactic in the book, although fortunately not the ones involving threats of bodily harm, but this one is my all-time most useful.

Showstoppers

In 1998 the UK National Audit Office produced a report into the £1.7 billion 1996 flotation of Railtrack, part of the highly complex and politically charged privatisation of British Rail. The situation was loaded with hidden agendas, but the NAO determined that "*Government's policy was to secure the sale of Railtrack within the lifetime of the then current parliament*" and the "*main objective for the flotation of Railtrack...was to secure the flotation as soon as reasonably practicable.*"

This still strikes me as faintly bizarre. Normally when you do something that big, you do it in order to achieve something. Yet the basic objective (or at least the only one the government admitted to in public) was to just do it. JFDI, as the chairman used to say.

But I don't think the NAO had misunderstood the objective. My job at the time was to write the prospectus, and my impression was that it would be issued come what may. The only possible showstopper was the fall of the government before the flotation could be accomplished. I used to check the news every morning to see if any more Tory MPs had kicked the bucket overnight. Actuarially speaking, the aging government was going to fall before the flotation could be accomplished, but Tory MPs are a tenacious bunch.

Nothing was allowed to stand in the way. Performance regimes not fully tested in circumstances of disruption? Make the train companies pay £75 million a year in access charge supplements to compensate. Shaky balance sheet? Write off £869 million of debt. Not sure if employees and the general public will buy shares? Add £47 million of incentives. Institutional investors don't see growth opportunities? Include the Thameslink 2000 project, reducing flotation proceeds by around £125 million. Directors still edgy about signing the prospectus? Promise the chairman a knighthood.

Measured against its simple objective, the privatisation was a resounding success. The government lasted just long enough to get Railtrack away, so the only identified showstopper didn't stop the show.

On the other hand, the subsequent history of the company was not exactly glorious: multi-fatality train crashes, multi-billion-pound cost overruns and a complete breakdown in the relationship with government that saw Railtrack put into administration in 2001.

Maybe identifying a few more showstoppers might not have been such a bad idea.

Accountability

There are two sorts of accountability. The one where someone makes themselves accountable for delivery of a project and the one where someone is held accountable after the fact for failure to deliver.

Ideally, the two sorts should go together – if you have accepted accountability for delivery then yes, you should be the one who takes the blame if delivery doesn't happen.

Problem is, if no one has accepted accountability for delivery in the first place, then:

    a. the chance of the project going pear-shaped just went up dramatically; and

b. blame will probably be sprayed around unfairly when it
    does go wrong

Lose-Lose.

What doesn't get measured …

In February 1995 State Rail, the public sector owner and operator of Sydney's rail network, let contracts to build a new underground rail line from Central, out past the airport to join the
East Hills line at a new station called Wolli Creek. The one
which caused most of the angst was the Stations Agreement.
This set out the terms on which the private sector would design,
construct, finance, lease, operate and maintain four stations,
including the ones at the domestic and international terminals.

The line opened in May 2000. The station contractor was in
trouble within a month and in receivership before the end of
November. The speed of the collapse is a measure of how badly
people got things wrong. (A big factor was the demand forecasts,
always an issue on transport projects. Even five years later,
patronage was about 30% of the forecast level.)

But the government had to reach a settlement with the contractor, because State Rail was in breach of its obligation under the
Stations Agreement to run at least a set number of trains an
hour through the stations. The obligation was undeliverable
with the infrastructure then available.

Not clever. How did they come to take on an undeliverable
obligation?

At the time, they simply didn't have systems more sophisticated
than someone with a clipboard on a station to check the number
of trains going through, and they didn't normally bother to do

even that. Now, of course, train delays are measured to the minute and cancellations are tracked electronically, but on-time running in the 1990s wasn't a thing people worried about so much. There was a timetable, which showed the right number of trains, but nobody had a handle on how many of those trains would actually run.

As it turned out, not enough.

So the promise was made without any supporting capacity behind the promise. Arguably with better management State Rail could have got a lot closer to running the right number of trains, but fundamentally it took on obligations that were just not deliverable.

Bad move.

PART 12

Cost overrun?

Does it count as a cost overrun on a PPP if the client doesn't pay for it?

The whole point of a PPP is to transfer risk to the private sector. Governments don't always do it very well. When Metronet went bust delivering the London Underground PPPs, the risk came straight back to the public sector.

But sometimes the risk transfer works as intended.

Owen Hayford, at infralegal.com.au, gives a number of Australian examples of cost overruns in a 2021 paper on improving PPPs. In each case it was the design and construct contractor that took the risk – and the hit.

- Southern Cross Railway Station: Leighton Holdings announced forecast losses of $122.6 million as a result of cost overruns under the D&C contract

- Waratah Train: Downer EDI announced losses totalling $440 million on the D&C contract
- Victorian Desalination Plant: in March 2012, Leighton Holdings announced that it expected to make a loss of $602 million on the D&C contract, after originally forecasting a profit of almost $300 million

The government client in each case would have suffered some additional cost because of the delays associated with the cost overruns, but quite likely within the planned contingency.

So where do these projects go in the statistics? As a successful project with no cost overruns, which is what it felt like to the client?

Or as a massive cost overrun, which is what it felt like to the D&C contractor?

Hitting the ground

Early in my career, when I had given up being employed as a lawyer but was still trying to work out what my new career actually was, I had just finished helping to set up a major infrastructure contract.

One of the directors called me in and asked if I would like to be the account executive.

Um.

I had no idea what an "account executive" did. It is possible my response was both hesitant and incoherent.

The director promptly decided to offer the job to somebody else, very much to my relief when I worked out the account executive was going to be responsible for managing the contract we had just set up. I didn't remotely have the skill set to manage a tier I contractor.

Looking back, it seems to me the height of incompetence that the company didn't have a contract manager in place and ready to go at the moment the contract was signed. Ideally, the contract manager should have been in place through the procurement phase, gaining familiarity with the contract and the stakeholders.

Those first weeks are critical in building the relationship between the parties, establishing patterns of behaviour that will colour the whole delivery period. The absence of a contract manager sends a strong message to the contractor that the client has no bloody idea.

And yes, that project was delivered late and over budget. What a surprise.

The affirming negative

Saying "no" is a core skill for project managers.

Unfortunately, when the demand comes in to achieve the impossible (and on a megaproject it's a when, not an if, this is the real world we're talking about) the Minister doesn't want to hear a negative.

A simple affirmative will satisfy the Minister in the moment but, since the demand is impossible, disaster will inevitably follow.

Survival demands that you disguise your negative as an affirmative. You don't agree to comply with the impossible demand (that's the negative bit) but you do agree to deliver something else (that's the affirmative bit). If you craft it carefully enough, the Minister will either fail to notice the difference or be convinced the change is their idea.

My best ever affirmative negative came on the Waratah train project, where an impatient premier announcing the procurement set a ridiculously short target to issue a request for expressions of interest on what turned out to be a $3.6 billion project. It was technically possible, but the request would have been rubbish, since we didn't yet know what the project was. Yes, we knew we were buying a lot of trains, but that usually comes with requirements for stabling and maintenance and traction supply upgrade and all sorts of other stuff and we hadn't worked that bit out yet. EOIs are used to create a shortlist of bidders and you *really* don't want to start shortlisting before you've figured out the deliverables.

A simple affirmative was no good. But a simple negative would have been no good either. At the time, it was politically important to be seen to be making progress with the project.

So we disguised the negative as an affirmative. We called for registrations of interest. "Registrations" not "expressions". We were going to need to issue a real EOI later, so we had to call this exercise something different, but it also had to be similar enough to appear to be complying with the premier's wishes.

Government had already decided to impose a local content requirement, so we invited anyone in the country interested in doing anything to do with delivering trains, such as manufacturing train seats or air-conditioning units, to register that interest. In return, any local business which registered would have its details supplied to the tenderers as a potential partner in meeting the local content requirement. They also received an invitation to attend a presentation with tea and biscuits and a harbour view.

If the premier noticed the difference between this and a standard EOI process, he didn't care. More than a hundred people showed up to the presentation, including all the local media.

Local businesses had a networking opportunity. The premier had his photo opportunity. The project team had an extra-time-to-prepare-for-the-real-EOI opportunity.

Happiness all round.

Saving time

The amount of time you lose by doing something stupid exceeds the time you thought you were going to save by doing it.

In 1991 the Australian Civil Aviation Authority commenced a tender process for the procurement of a new air traffic control system. They were really keen to get it done quickly and produced a timetable showing contract signature in 12 months.

A procurement consultant was engaged to comment on the process before the Civil Aviation Authority went to market. Not surprisingly, he advised the twelve-month timetable was *"extremely tight if not impossible"*.

The advice was not only ignored, the process was subsequently revised to reduce the timetable even further. Not the most intelligent decision.

In the haste to award a contract, the Authority made mistake after mistake. The first appointment of a preferred bidder was set aside after a challenge and the last part of the process had to be re-run with the two shortlisted bidders.

Far from having learned its lesson, the Authority was later found to have failed to evaluate the new tenders in accordance with the methodology set out in the request for tenders and to have accepted an out-of-time change from one of the bidders – elementary process errors that could easily have been avoided.

A contract was not signed until almost two years after the original target date and the Authority was still unsuccessfully defending its actions in the Federal Court three years later.

Abandoning a good strategy

What is it about light rail and utilities relocation? We all know it's difficult. The older the city, the more unrecorded pits and pipework. But that gets taken into account, doesn't it?

Apparently not. The ill-fated Edinburgh Tram project had an excellent utilities strategy. All the utilities were to be diverted prior to letting the tramline contract, giving the construction contractor a free run, clear of the usual delays associated with utilities relocation. Really impressive.

Once work got underway on the relocation, it was discovered the utility company records were inaccurate (surprise!) and there was a ton of redundant apparatus and other obstructions down there (more surprise!!), so the relocation works were delayed (even more surprise!!!).

At this point, presumably spooked by the unexpected arrival of the totally foreseeable, they threw their excellent utilities relocation strategy out of the window.

They went ahead and let the construction contract anyway, without putting in place adequate provision for coping with parallel construction and relocation works. Oh, and the design was also supposed to be finished before the contract was let, but that didn't happen either.

The upshot? The tramline was three years late and some 40% over budget, despite being a couple of kilometres shorter than originally planned.

Not a happy ending.

Corporate reorganisations

Corporate reorganisations have a lot to answer for. In 2011 the UK Department for Transport restructured itself into a major problem with train franchising.

The process of letting franchises to train operating companies had previously come under a single Director-General. The new organisation was a matrix structure where the Domestic Group did franchising policy, the Major Projects and London Group did the letting of the franchise contracts and the Corporate Group had the refranchising finance expertise.

The intention in the new structure was that the Domestic Group would provide the Senior Responsible Owner for the refranchising programme as a whole. As each franchise moved to the procurement phase, responsibility for that franchise would be passed to the Major Projects and London Group.

So when the InterCity West Coast franchise moved into the procurement phase in January 2012, responsibility for it was thrown over the new silo wall into the Major Projects Group. Nobody caught it.

And nobody noticed. The Contract Award Committee, which was supposedly providing assurance on the procurement process, failed to meet at all between 16 January and 19 June, which was a bit limiting on the assurance front.

The absence of an SRO was picked up by a Gateway review, and in April the project was allocated an SRO from the Major Projects Group. Even then, the role was at a refranchising programme level rather than at project level. There was no defined relationship between the SRO and the Contract Award Committee and no clear reporting line for the project team leader.

The procurement process was a failure. The preferred bidder award was challenged by the losing bidder and the whole process had to be cancelled.

The decision that cruelled the entire procurement was to use an unfit-for-purpose financial model to calculate the amounts of subordinated loan facilities and to use it in a way contrary to guidelines given to the bidders.

You probably can't blame that entirely on the corporate reorganisation. But I don't think it's a coincidence that the fatal decision emerged in the period before April 2012, the hiatus where there was no SRO and no clear reporting line.

Things get messy when no-one is in charge.

Clear objectives

If the objectives haven't been properly captured, you could end up building the wrong thing.

In the autumn of 2004, after a sweltering summer, the NSW government was under siege by commuters fed up with sweating their way to and from work on unairconditioned trains.

There were three more summers until the next election, and the issue was clearly going to run and run. They decided to replace the old unairconditioned passenger rolling stock with new trains that would become known as Waratahs.

As the procurement process progressed, revenue shortfalls started to put serious pressure on the state budget.

A Treasury official, keen to keep the books balanced, correctly pointed out that if the new trains were also not airconditioned, the replacement would be significantly cheaper. There was, after all, nothing in the project objectives to make airconditioning necessary.

He did leave the room alive, but it was touch and go for a minute there.

A requirement for the new trains to be airconditioned was promptly added to the formal objectives.

Where did the lake go?

It doesn't matter how much advance work you put in on the geotech if you then dig in the wrong place.

Lake Peigneur is a 60-metre-deep picturesque lake in Louisiana. It used to be only 3 metres deep.

In 1980 Texaco, searching for oil, leased a drilling platform in the lake. A 14-inch drill bit went down 360m without incident. And kept going. At about 370m it penetrated a mine tunnel in the salt dome below the lake. That wasn't supposed to happen.

The lake water drained into the mine, dissolving the salt pillars supporting the mine ceilings and collapsing the mine. The Delcambre Canal that normally formed the outflow of the lake temporarily reversed its flow and created a fifty-metre-high waterfall into the new lakebed. The vortex of water into the mine sucked down the drilling rig, eleven barges, a tugboat and 65 hectares of land.

Miraculously, the workers on the rig made it to shore and the 55 miners were safely evacuated. Three dogs died: the only fatalities.

So much destruction was caused that the Mine Safety and Health Administration was unable to allocate blame for the disaster: they could not determine whether Texaco was drilling in the wrong place or if the mine's maps were inaccurate.

That didn't prevent the inevitable lawsuits. The mine owner received $32 million in an out-of-court settlement and the landowner a further $12.8 million.

Not the world's most successful attempt at drilling for oil.

What are the odds?

Can you believe a lawyer about the prospects of success in litigation?

I'm always particularly wary if I'm told the probability of success is 70%. It's a nice round number, which is sufficiently positive to encourage the client to proceed with fat-fee litigation, but the implied 30% chance of losing is big enough that the lawyer won't look too stupid if you do in fact lose. I'm not convinced the figure bears any relationship to the actual chances of success.

When the UK Nuclear Decommissioning Authority was sued by the losing bidder over its award of the £6.2 billion Magnox contract in 2014, it sought Counsel's advice on the prospects of successfully defending the case. Counsel was initially "*cautiously optimistic*", but as preparation for the hearing progressed, he changed his mind.

"I remain of the view that the case could go either way ... However, if I was asked to say whether I thought we were more likely to win or lose, I would say lose, albeit only by a narrow margin."

I may be wary of lawyers inflating your chances of winning a court case, but when even your own Counsel thinks you're probably going to lose, it's definitely time to make an offer to settle.

The NDA refused, apparently thinking an offer of settlement would encourage others to sue, so, in their own words, *"[E]ven if the liability case were lost, this might be the better outcome."* An approach even their own Head of Procurement described as "*ridiculously uncommercial*".

The case was fought and lost.

Should have listened to the lawyer.

What you know for sure

"*It ain't what you don't know that gets you into trouble. It's what you know for sure that just ain't so.*" Mark Twain knew what he was talking about.

One of the key factors in the demise of Railtrack (predecessor of Network Rail) was that in the late 90s it entered into a fixed price contract to allow Virgin Trains to run additional services on the West Coast Main Line.

There was already a £2 billion upgrade of the line getting underway, funded by government through regulated access charges. But providing the paths for the extra Virgin Trains services would require additional infrastructure.

A fixed price was negotiated for the additional works, with Railtrack working very hard to ensure its cost estimates were reliable.

The estimates for the original upgrade didn't get the same level of attention. It didn't seem necessary, because works under the regulated regime were essentially risk free. Railtrack knew that for sure.

To get the markets to accept Railtrack's privatisation in 1996, in the face of visceral opposition from what would in 1997 become Tony Blair's government, Railtrack had been given a sweetheart deal on infrastructure works. Provided it carried out works efficiently (hard to disprove), there was no comeback if costs went up. No worries if the estimates were a bit out, government would have to pay up if it wanted the works done.

It hadn't occurred to anyone that costs might go up so far the government would no longer be prepared to continue funding the works at all. It occurred to everyone when the forecast went from £2 billion to £4 billion. And kept climbing.

Railtrack had been right that there would be no direct comeback if costs went up. If government wasn't prepared to fund the work, Railtrack didn't have to do it. But they were wrong to have assumed government would keep shovelling in the money.

The consequence of that miscalculation was dire. Without the upgrade, the additional works would not be enough to deliver the train paths promised to Virgin Trains. Railtrack couldn't fund the upgrade works itself, but neither could it fund the horrendous break costs for the Virgin contract.

A deal which had been hailed as proof of claims that privatisation would allow new growth on the railway was instead opening up a black hole in the Railtrack accounts. The company was finally forced into administration in 2001.

Government did eventually fund Network Rail to complete the upgrade, at a cost thought to be around £8 billion. Too late for Railtrack.

Delusional scheduling

The problem with an insanely optimistic delivery date is not that it's not achievable, it's that people insist on pretending it *is*.

It happens on major infrastructure procurements too often. A Minister announces a shiny new project and an impossible target date for signing a contract. The project director, perhaps owing to the absence of a convenient bike shed to take the Minister behind for an informal re-education process, spinelessly produces a schedule showing contract signature on the desired date.

The project director knows perfectly well the date is unachievable, but figures that somewhere down the line, the Minister will be forced to realise it's impossible and will grant an extension of time. The impossible schedule is just an interim measure.

But this way lies disaster. The only way to make the interim schedule match an impossible deadline is to cut short the planning and consultation phase and shove out a request for tenders before you've done the work to get it right.

And yes, the bidder response to the request for tenders will be sufficiently negative to convince the Minister the deadline has to be extended. But then you start from where you are. You never get to do the planning and consultation you should have done.

The extension of time doesn't come with recognition that your specification was hopelessly underdeveloped and should never have been put out to tender in its present form. Or that there is still too much uncertainty around key risks to be able to determine a sensible risk allocation. The planning phase is officially over, and you haven't done the planning.

The probability of bringing the project in on time and on budget? Zero.

The screwdriver error

Most screwdrivers are vastly over-engineered for turning screws, but good hardware shops don't sell the cheap thin ones – customers complain because the screwdrivers break when used to lever the lids off paint cans. (Well, duh, that's what screwdrivers are for, isn't it?)

There's a lesson there for organisational change exercises – people may be doing important things that aren't apparent from the job title or even the official job description.

I recall one clerk at British Rail, in the section which became Railtrack back in the 1990s. One of his tasks, which occupied him no more than half a day a week, was to check freight invoices. After the privatisation, the other four-and-a-half days' worth of his activity became redundant and so did he.

Alas, after his departure the task of checking freight invoices was overlooked and not re-allocated. Several years later, the organisation woke up to the fact that it had underbilled freight operators by about £2 million, probably irrecoverable.

The modularity principle

In their book, *How Big Things Get Done*, Flyvbjerg and Gardner identify modular design as the key distinction between nuclear power projects and solar power projects. A nuclear power station is a big and complicated one-off, but a large solar array is created by clipping small solar arrays together: simpler to do in the first place and you learn to do it better and better as the project progresses. No prizes for guessing which type of project is likelier to come in on time and on budget.

Modularity doesn't just apply to design. Perhaps surprisingly, the concept translates well into contract administration.

Big projects aren't just about the engineering. With any large contract, there will be a correspondingly large volume of contractual interactions. Invoicing is a classic. Monthly payments, milestone payments, variation payments. On a five-year contract there will be dozens, perhaps hundreds of invoices from the contractor.

Modularity is your friend. Instead of just letting the accounts department get on with it, treat the first invoice as a pilot. Get client and contractor together and critique the process. Was the invoice submitted on time? Was the format helpful? Did the data drop out of the system or was manual processing required? Was the right information demanded and submitted? Were there any surprises? How long did the approvals process take? How long before the money appeared in the contractor's bank account? Can you do the invoicing and payment process better/faster/cheaper? Put the results of the review into the second invoice and repeat until the process is as frictionless as possible.

It is not uncommon to find parties still arguing years into the contract about the evidence required to support invoices, and there is nothing like arguments about payment to destroy trust.

A bit of effort upfront to create repeatable, frictionless processes can transform contract administration.

When you're in a hole …

The construction of Berlin Brandenburg International Airport is notorious for the size of the dent it put in Germany's reputation for efficiency. Construction began in 2006 with completion planned for 2011, five years away. It took fourteen.

A lot of things went wrong, but the ever-changing contracting strategy was particularly noteworthy. The government-owned company responsible for building the airport just couldn't get it right.

Their first attempt at a contract saw them with two consortia bidding to construct the airport and operate it for 50 years. The 1998 award of the contract was annulled when the losing bidder sued: bias was found in part of the evaluation.

The second attempt saw the two consortia get together to put forward a joint proposal. This was rejected by the agency board in 2003. It paid the consortia some €50 million for their efforts and scrapped the privatisation plan altogether.

The third attempt saw the agency decide to have a single general contractor be responsible for the detailed design, construction planning and construction. This attempt also failed, when all four bidders for the contract bid the price at about €400 million above the agency's €630 million estimate.

The fourth attempt saw the project broken into seven separate contracts which were put out to tender. Lo and behold, the bids that came back added up to about €1.1 billion, fairly adjacent to the price rejected for the single contract. Possibly at this point someone should have introduced them to the notion that when you're in a hole, you should stop digging. Nobody did.

The fifth attempt instead broke the project up into even more contracts, moving from fixed price to unit rates in an attempt to save money by reducing contractor contingency. Instead of only overseeing one contractor who would also be responsible for the detailed design, the agency now took design responsibility and had to manage interfaces with around 35 contractors. Alas, the agency simply did not have the competence to manage a project on this basis. It didn't help that they allowed construction to commence before the detailed design was finished.

The airport finally opened in 2020 at a cost somewhere north of €6.5 billion.

The evaluation report

When settling an evaluation report, pretend you are the losing bidder. Even better, the losing bidder's lawyer. What would they think if they read it?

Why was the other bidder preferred? Does this gap in the chain of logic indicate an evaluation error? How could someone with our terrific score for widget quality possibly fail to win the contract? Why is there no mention of the key features of our proposal? What evidence is there for saying our bid is more risky? Are the evaluation criteria exactly as they were published?

Although the bidders won't normally see the evaluation report, you will have to debrief the losers and explain why they lost. If the report doesn't contain an explanation you would feel comfortable presenting to the losing bidder, the report is not adequate.

Bid teams are the sales division of contracting firms and if there is one thing the best sales professionals are good at it is reading body language. If you aren't convinced by the explanation, they won't be either. And if they aren't convinced, they may challenge the decision.

In the 2012 InterCity West Coast franchise case, the project team had decided not to release to bidders the model used for calculation of the subordinated loan facility (an important factor in the evaluation) because they were concerned it could be challenged. That was not, of course, what they told the bidders, but the message got across anyway.

The subsequent Laidlaw Inquiry quoted a note made by the losing bidder of a meeting with the project director:

"Eventually, after MUCH more debate and probing, I believe I got to the heart of the matter. Whilst it was presented less obviously, he was effectively telling me that they COULD NOT contemplate releasing their stress-test model because it is very basic and would be open to challenge."

You need not and should not allow *"debate and probing"* in a debrief meeting. The decision has been made, it is not up for debate.

Nonetheless, the losing bidder is entitled to know the reasons for the decision and it is very much in your interests that they should find the reasons convincing.

So read the draft evaluation report as if you were the losing bidder's lawyer, looking for grounds to challenge the decision. And don't give them any.

Contracts written with a chisel are shorter

Which was demonstrated in 324BC, in the first private financing deal in recorded history. Thank you to Professor Eric Csapo and Anthony Alexander of Sydney University for drawing this to my attention.

The public sector organisation involved was the City of Piraeus in Greece. Piraeus was then as now a major transport hub, being the port city for Athens, but the deal was not a transport PFI. It involved the old theatre.

By 324BC some Greek cities already had the famous stone theatres that are still visible in Greece today. Piraeus either didn't want or more likely couldn't afford one, so it followed the older theatre model where the seating and stage were principally wooden. They couldn't be left in place as the wood deteriorated very quickly in the Greek climate, so they had to be taken away and stored until the next religious festival or commercial event. This was a nuisance and an expense for Piraeus, so the City put the concession out to tender.

The tender was won by a consortium of four citizens who paid a large capital sum for the right to supply the seating and to take a share of the gate. That is, the private sector financiers took demand risk. They were required to keep the theatre area to a certain standard: if they failed, the City had step-in rights. There were asset condition standards on handback. The terms of the contract had to be published. Very much the typical PFI.

Okay, publication was on a stone tablet in the agora rather than on the internet, but hey, that was what they had.

The City even awarded a crown, the moral equivalent of an

OBE, to one Theaios, it seems because he got a better deal out of the consortium than the public sector comparator.

Although history does not record the outcome, the consortium probably made a mint. In the days before television and printed books, the theatre had a virtual monopoly on entertainment: think religious festivals, think the plays of Aeschylus, Sophocles and Euripides. Basically, it was like having exclusive rights to Shakespeare, Taylor Swift and the Pope.

So far as we know, later stone theatres were financed by the State, which suggests the State got fed up with the private sector pocketing all the proceeds.

Nothing new under the sun.

WANT MORE?

Find more mini-case studies on LinkedIn
https://www.linkedin.com/in/louisehart750/
Or read my book
*Procuring Successful Mega-Projects: How to Establish Major
Government Contracts Without Ending up in Court*

ACKNOWLEDGMENTS

Most of the case studies in this book were originally written as posts on LinkedIn. Many people posted comments and reactions in response, which have been enormously important in sustaining my enthusiasm, correcting my errors, providing me with ideas and improving my writing skills. My thanks to you all.

Special thanks to: my beta readers, Sallie Mason, Helen Murray and Dominique Tubier, who suggested numerous improvements; my brother Stephen and his wife Pamela, who kindly lent experience from their own publishing imprint, *Improbable-Fictions.com*; and to my husband Anthony, who puts up with a lot.

A NOTE ON SOURCES

The case studies have been derived from many sources, including my own experience, experiences shared by others, media reports, inquiry reports, books, Wikipedia and other internet resources. I have often included informal attributions within the posts, knowing a simple internet search will lead you to the source if you're interested to look further.

Some projects deliver more lessons than others. There are more than 200 mini case studies in this book. About a quarter of them were taken from just six projects.

The case studies in relation to two of these, the Waratah train procurement and the Railtrack privatisation, are taken directly from my own experience. For the remaining four, you may enjoy a deeper dive into some of the original sources. Further reading is suggested below.

I have also included a list of the books referred to in the posts, plus a couple of extras that are well worth a read.

THE BIG SIX

Edinburgh Tram project

The Inquiry into the Edinburgh Tram project took nine years to deliver its report. If that sounds like a long time, well, there was a lot to write about. The idea was to build a tramline from the airport to Newhaven. Contracts were signed in 2008, and the line was expected to open in 2011. After a chapter of errors (or, if we follow the Inquiry report, 25 chapters of errors) it eventually opened in 2014, way over budget, with the line ending not at Newhaven but at York Place, about two kilometres short of the intended destination.

For anyone who cares about learning from past projects, there are many, many lessons in the Inquiry report:

- Lord Hardie, *Edinburgh Tram Inquiry Report*, www.edinburghtraminquiry.org, August 2023

Case studies: A CEO in conflict; Abandoning a good strategy; Assume a can opener; Corporate conflict; If you see something, say something; Lessons delayed; Pockling the spreadsheet; Self-incrimination by email; SRO missing in action; The nuclear option; When is a fixed price not a fixed price?; Working off the wrong baseline.

Ferry fiasco (Scotland)

That's the heading for the Wikipedia entry on this project. As at March 2025, only the *Glen Sannox*, one of the two ferries ordered for delivery in 2018, had made it into service. The *Glen Rosa* is still a work in progress. The contract was awarded, on the strength of a less-than-best-practice evaluation process, to a local shipyard, which promptly went bust and was nationalised by the Scottish government. The tale of woe can be followed through the Scottish media. There have also been various formal reports relating to the project. The two most useful are:

- Audit Scotland, *New Vessels for the Clyde and Hebrides: Arrangements to deliver vessels 801 and 802,* 23 March 2022
- Public Audit Committee of the Scottish Parliament, SP Paper 344, *New Vessels for the Clyde and Hebrides: Arrangements to deliver vessels 801 and 802*, 23 March 2023.

Case studies: Breaking the whisky bottle; Consulting the operator; Local industry preference; Mandatory criteria; Milestone selection; More haste, less speed; Risk allocated to a contractor ceases to exist; Should you evaluate mandatory criteria first?; What is it about ferries?

InterCity West Coast franchise

In 2012 the UK Department for Transport embarked on a tender process for the InterCity West Coast franchise, one of the largest passenger rail franchises in the country. A preferred bidder was selected, but the runner-up, Virgin Trains, brought an action for judicial review before the £5.5 billion contract could be signed. Various deficiencies were identified, including fundamental failures of governance. The competition was cancelled, meaning Virgin Trains, as the incumbent, secured a two year extension of its franchise to enable the competition to be re-run.

The inquiries into the failed competition were thorough and provide an unusually close and fascinating look at the workings of a government tender process:

- *Report of the Laidlaw Inquiry,* 6 December 2012 (HC 809)
- *Report by the Comptroller and Auditor General,* 7 December 2012 (HC796)
- *The Brown Review of the Rail Franchising Programme,* January 2013 (Cm 8526)
- *Transport Committee Eighth Report of Session 2012–13* (HC 537)

Case studies: The wrong project governance framework; Buying time; Corporate reorganisations; How many megaprojects is too many?; In-house expert v. External expert; Lacking assurance; The clarity of hindsight; The hidden costs of getting it wrong; The evaluation report; To escalate or not to escalate; Treating bidders badly.

Queensland Health payroll debacle

The introduction of a new payroll system for Queensland Health in March 2010 was described by the subsequent Commission of Inquiry as a *"catastrophic failure"*. Thousands of staff received the wrong amount, or nothing at all. It took months to develop a functioning payroll system, with more than 400 additional payroll staff engaged to enable Queensland Health to pay its employees. The case can be used to draw many different lessons, since there were many different mistakes. Particularly notable was the complexity of the project requirements: the payroll system had to support 85,000 staff employed under two different pieces of legislation, covered by 12 industrial awards and impacted by 6 different industrial agreements, which between them created over 200 allowances and up to 24,000 combinations of pay. Government failed either to simplify the project requirements or to implement any effective management of the risks imported by such complexity.

Further detail is available in the following reports:

- The Hon Richard Chesterman AO RFD QC, *Queensland Health Payroll System Commission of Inquiry Report*, 31 July 2013
- KPMG, *Queensland Health: Review of the Queensland Health Payroll System*, 31 May 2012
- Auditor-General of Queensland, Report to Parliament No.7 for 2010, *"Information systems governance and control, including the Queensland Health Implementation of Continuity Project"*, Finance and Compliance Audits, June 2010

Case studies: Doing the reference check; Going the wrong way; Payroll problems; Self-incrimination by email; What is it about confidential information?

Railtrack privatisation

Railtrack was created as part of the controversial privatisation of British Rail in the mid-1990s. It was owner and operator of the railway infrastructure formerly owned by British Rail and also my employer from 1995 to 2002. The privatisation process was unimpeachable: no action was ever brought under the prospectus even though the later abrupt descent into administration occurred within the applicable statute of limitations. (I was quite proud of that: my first job at Railtrack was to write the prospectus.) Success had a price, measured in the compromises made to secure the flotation. The subsequent history of Railtrack was short and inglorious. The delivery of Stage 1 of the Channel Tunnel Rail Link (CTRL) on time and on budget, perhaps its one genuinely impressive achievement, was airbrushed out of history when its interests in CTRL were divested as part of the 2002 settlement with the UK government following the administration.

Case studies (including CTRL project): The core business of a railway is ...?; Assessing the evidence; Cost estimates way below outturn. Why?; Death by a thousand cuts; How to be commercial; Shovel-ready projects; Showstoppers; The menace of silo budgets; The sex life of badgers; What you know for sure; The screwdriver error.

Waratah trains PPP

I was employed by Rail Corporation NSW from April 2004 as Project Director for the establishment of the $3.6 billion Public Private Partnership which delivered 78 passenger trains to the Sydney network. I handed over responsibility for delivery to my successor once contracts were signed at the end of 2006 and worked on various other projects for RailCorp until I left in January 2014. All opinions and observations in this book are mine and should not be attributed to my employer.

Case studies: A unique PPP?; Clear objectives; Client decision or bidder decision?; Media massage; Open for business?; Spotting the difference; Tadgell's Bluebell; The affirming negative; The world moves on.

SELECT BIBLIOGRAPHY

Paul Carroll and Chunka Mui, *Billion Dollar Lessons: What You Can Learn from the Most Inexcusable Business Failures of the Last 25 Years*, Portfolio, 2008.

Henrico Dolfing, *Insights*, www.henricodolfing.com

Annie Duke, *Quit: The Power of Knowing When to Walk Away*, Ebury Digital, 2022.

Miles O. Frank, *How to Get your Point Across in 30 Seconds or Less*, Gallery Books, 1990.

Bent Flyvbjerg and Dan Gardner, *How Big Things Get Done: The Surprising Factors That Determine the Fate of Every Project, from Home Renovations to Space Exploration and Everything in Between*, Macmillan 2023.

Sir Francis Fox, *Sixty-Three Years of Engineering*, London, 1924.

David Gration et al., *White Elephant Stampede: Case Studies in Policy and Project Management Failures,* Connor Court Publishing Pty Ltd, 2022.

Empire State Inc., *The Empire State*, New York: Publicity Association, 1931

Louise Hart, *Procuring Successful Mega-Projects: How to Establish Major Government Contracts Without Ending up in Court*, Routledge, 2015.

Owen Hayford, *Insights*, www.infralegal.com.au

Andrew Hopkins, *Failure to Learn: The BP Texas City Refinery Disaster*, CCH Australia Reprint Edition, 2015.

Anthony King and Ivor Crewe, *The Blunders of our Governments*, Oneworld Publications, 2013.

Brian Klass, *Fluke: Chance, Chaos and Why Everything We Do Matters*, Scribner Book Company, 2024.

Stephen Pile, *The Return of Heroic Failures*, Martin Secker & Warburg Ltd, 1988.

James Reason, *Managing the Risks of Organizational Accidents*, Ashgate Publishing Limited, 1997.

Nevil Shute, *Ruined City*, Cassell, 1938.

Jessica Pooi Sun Siva and Thayaparan Gajendran, *Power in Megaproject Decision-making: A Governmentality Approach (Spon Research)*, Routledge, 2024.

Ozan Varol, *Think Like a Rocket Scientist: Simple Strategies for Giant Leaps in Work and Life*, WH Allen 2021.

Christian Wolmar, *On the Wrong Line: How Ideology and Incompetence Wrecked Britain's Railways*, Kernsing Publishing, 2012 or Aurum Press 2005, revised and updated from the original, *Broken Rails*, Aurum Press 2001.

ABOUT THE AUTHOR

Louise Hart has been around megaprojects in the UK and Australia for more than 35 years, originally as a lawyer, then as a transaction manager, a railway executive, a megaproject director, and now as an independent adviser based in Sydney.

In addition to her client work in major projects and infrastructure procurement, Louise is a member of the Dispute Review Board Foundation and currently serves on dispute boards in the transport and social housing sectors.

Her first book, *Procuring Successful Megaprojects: How to Establish Major Government Contracts Without Ending up in Court*, has become a bible for project professionals.

www.ingramcontent.com/pod-product-compliance
Lightning Source LLC
Chambersburg PA
CBHW071341210326
41597CB00015B/1532

*9 7 8 1 7 6 4 0 3 3 8 0 0 *